NANOTECHNOLOGY SCIENCE AND TECHNOLOGY

NANOTECHNOLOGY

PRINCIPLES, APPLICATIONS AND ETHICAL CONSIDERATIONS

Nanotechnology Science and Technology

Additional books in this series can be found on Nova's website under the Series tab.

Additional e-books in this series can be found on Nova's website under the e-books tab.

NANOTECHNOLOGY SCIENCE AND TECHNOLOGY

NANOTECHNOLOGY

PRINCIPLES, APPLICATIONS AND ETHICAL CONSIDERATIONS

PARESH CHANDRA RAY
EDITOR

nova
science publishers
New York

Library of Congress Cataloging-in-Publication Data

ISBN: 978-1-53613-889-4

Published by Nova Science Publishers, Inc. † New York

Contents

PREFACE

The emerging science of nanotechnology which deals with the understanding of the fundamental physics, chemistry, biology, material science and technology of nanometer scale objects has become a central pillar for the next generation medical challenges such as developing tiny nanodevices, as well as for food technology. There is no doubt that the development of emerging nano-technology based nano-medicinc, nano-devices for diagnostic and therapy, drug delivery systems and other applications are fast growing research areas for chcmistry, biology, physics, medicine and different disciplines of engineering. On the other hand, due to the possible use in human health and food technology, same emerging technology might raise new ethical issues and therefore requires careful analysis of ethical aspects. To summarize the recent growth, the first volume in the, *Nanotechnology: Principles, Applications and Ethical Considerations*, series discusses the basic science behind the emerging technology, which is necessary to understand how these tiny materials can be used in our daily life. This book provides state-of-the-art overview of this rapidly-expanding and fascinating field from the molecular level to possible applications in medical field. It contains eight chapters written by the world leaders in this area, covering from basic science to possible device design which can have immense applications in our society. This book is unique in its design and content, providing depth of science, which will help the reader understand the benefits and limitations of nanotechnology as well as its ethical and social implications. I believe that the readers will be very pleased to read the wide range of start-of-the art nanotechnology application for nano-medicine and food technology. We thank all the expert scientists for their contributions and Nova Science Publishers, Inc. for printing and timely publication of the book for the future readers.

The first chapter entitled "Synthesis and Characterization of Iron Oxide-Gold Shell Nanoparticles," discusses the advancements in the design of iron oxide-gold core-shell nanoparticles using different synthetic route. The optical properties of nanomaterials that depend on the size of core and shell as well as on the geometry of the shell can be used to elucidate the structure-property relationship, which has been explained. Huang et al., from the Department of Chemistry, University of Memphis, Memphis, TN, USA, discusses the current advancement on the start-of-the-art design for core-sell nanoparticles. Authors are also discussed computational studies on growth mechanism for different nano surfaces.

The second chapter entitled "Multitudinous heteroatom doped luminescent carbon dots: A Versatile Nanoscopic Molecular Vehicle for Targeted Anticancer Drug Delivery," by Dr. Pramanik et al. from USA reports the synthetic strategies that has been used for the development of different multitudinous heteroatom doped luminescent carbon dots of various carbon dots. This chapter discusses the mechanism for carbon dots based anticancer drug delivery systems. It also discussed the pathway for intracellular uptake, pH-triggered drug release and monitoring clearance. At the end, they have concluded that carbon dots based anticancer drug delivery systems have several advantageous including low-toxicity, favorable biocompatibility, excellent water solubility and super-high drug loading capacities.

In the third chapter entitled, "Engineered Plasmonic Nanostructure for Biological Sensing," by Prof. Singh from USA reviews the recent progress in the various ways to probe the special properties of plasmonic nanostructure for biological diagnostics and therapeutics. Initially, the chapter deals with a brief survey of the available ways to prepare plasmonic nanostructures and associated advantages and disadvantages. After that, it discusses some of the pioneering methods to alter the surface properties by loading functional groups for specific biological sensing and high throughput performance. Current chapter also discusses about various highly sensitive spectroscopic techniques such as colorimetric, fluorescence, Raman and hyper-Rayleigh scattering spectroscopic properties of plasmonic nanostructures as probes to identify the targeted biological pathogens. At the end, in the future perspective section, they have indicated that the use of plasmonic nanostructures bring many advantages in comparison with other actual alternatives.

Chapter four entitled, "Nanoparticles for Photothermal Therapy in Cancer," by Vangara et al., from USA discusses about the different varieties of theranostic nanoparticle and their contributions to photothermal therapy in cancer. The current chapter discusses the use of nanoparticles, nanoshell, nanorods and nanowires and different inorganic materials for cancer therapy. It also discusses photodynamic therapy with reactive oxygen species generation, fluorescence imaging, photoacoustic imaging and optical At the end, the authors concluded that continued research on the development of multifunctional material is very important before it can be used in clinical settings.

In the fifth chapter entitled, "Iron Oxide Nanoparticles for Magnetic Resonance Imaging with Improved Sensitivity," by Dr. Zhao et al. from USA, discuss about the recent efforts to develop IONPs based contrast agents to improve the sensitivity of MRI. It reviewed the development of IONPs to increase transverse relaxivity as T_2-weighted MRI contrast agents. This chapter discusses the ability of IONPs to self-assemble and exhibit T_1-T_2 switching (sharp contrast transition) at different biological environment (i.e., pH) further enhances the IONPs sensitivity. At the end, authors concluded that the development of IONPs with high sensitivity will lead to earlier diagnosis of disease and help with designing treatment plan.

Chapter six, entitled "Multicolor Carbon Dots for Bioimaging," by Dr. Salma et al. from USA, discussed about the recent development of multicolor carbon dots based nanotechnology, which has been explored for bioimaging of virus and bacteria, cancer cells and brain cells. Current chapter discusses the formation mechanism of carbon dots from molecular precursors. It also discusses the cytotoxicity of carbon dots, which is important before it can be used for nanomedicine. In the end, authors discussed a great deal about the development carbon dots, which are expected to be excellent materials for *in vivo* and *in vitro* applications, because of their good aqueous solubility, bright photoluminescent characteristics, photochemical stability, low toxicity as well as biocompatibility.

Chapter seven, entitled "Nanoceuticals: Potential Application for Mitigating Nutritional Deficiencies," by Marotta et al., from Italy provides an insight into applications of nanotechnology in the field of nutrition and will shortly depict its advantages, possible unfavorable outcomes and solutions. This chapter discussed an insight into the different applications of nanotechnology in the food sector, neutraceuticals to nanoceuticals, mechanism of action of nanoceuticals and potential applications of nanoceuticals in overcoming nutritional deficiencies. At the end, they have concluded that utilization of nanoceuticals may enhance the absorption of supplements in comparison to traditional dietary supplements and they have a very high potential to mitigate nutritional deficiency.

Chapter eight, entitled "Ethical Consideration on Nanotechnology," by Ferreira et al. from Portugal discusses nanotechnology in terms of ethics. Current chapter discussed how the nanotechnology applications must satisfy ethics and consider the effects over the society and values. At the end, authors concluded that the ethical issues related with the themes "nanotechnology and environment" and "nanotechnology in the workplace" are extremely critical, and need to be well understood before nanotechnology can be used in daily life applications.

The current book provides foundation of a wide variety of nanotechnology approaches, which is invaluable for established scientists, junior researchers, and students involved in basic science, material science, medical filed and different disciplines of engineering. I hope that all the readers will be as excited as I am with the broad range of coverage on design principles, possible applications and ethical considerations of nanotechnology, which is an exciting new technology for promising applications for our society. We would value feedback from all readers of this book. Your comments are very important for us to improve the next edition. So please feel free to e-mail your suggestion to me via e-mail: paresh.c.ray@jsums.edu. Thank you for reading.

Paresh Chandra Ray, 4/21/2018

In: Nanotechnology
Editor: Paresh Chandra Ray
ISBN: 978-1-53613-889-4

Chapter 1

SYNTHESIS AND CHARACTERIZATION OF IRON OXIDE-GOLD CORE-SHELL NANOPARTICLES

Elyahb Allie Kwizera, Elise Chaffin, Yongmei Wang and Xiaohua Huang*
Department of Chemistry, The University of Memphis,
Memphis, Tennessee, US

ABSTRACT

With advancement of nanoscience and technology, many studies have been focusing on multicomponent nanostructures, especially hybrid nanomaterials that possess dual magnetic and plasmonic properties. Due to their dual properties, these nanomaterials are of great interest in many disciplines including science, engineering, and biomedical field. Currently, the most commonly used magnetic core is iron oxide and metal shell is gold. Great progress has been made in synthesis and characterization of these nanoparticles during the past two decades. This

* Corresponding Author Email: xhuang4@memphis.edu.

chapter will describe different ways in synthesis of iron oxide-gold core-sell nanoparticles followed by optical properties that depends on the size of core and shell as well as the geometry of the shell.

Keywords: iron oxide, gold, core-shell nanoparticles, synthesis, optical properties

INTRODUCTION

Iron oxide nanoparticles (IO NPs) are a major type of magnetic nanoparticles. Their chemical composition is magnetite (Fe_3O_4) or maghemite (γ-Fe_2O_3). Their magnetic moment aligns in the direction of external magnetic field, which leads to the concentration of the external magnetic flux density [1]. IO NPs are superparamagnetic when the sizes are small (lower than 30 nm) [2]. The superparamagnetic IO NPs avoids residual magnetization induced aggregation associated with ferromagnetic NPs. Because of the attractive magnetic properties, IO NPs are useful for many applications such as magnetic resonance imaging (MRI), biological separation, data storage, spintronics, and hyperthermia treatment of cancer [3-6].

To expand the function of IO NPs, research has been focusing on coating the IO NPs with a layer of Au during the past two decades. Coating the IO NPs with a gold (Au) layer will form a compact and hybrid core-shell platform that combines the magnetic properties of the IO core and the optical properties of the metal shell. Hybrid NPs with combined magnetic and optical properties can serve as magnetic NPs or optical NPs or a new nanoplatform. Thus, they are much more powerful than the single platform and can be used in a broad range of areas such as magnetic resonance imaging, optical imaging, biological separation, molecular/cellular detection, and disease treatment [7-30].

Driven by the powerful functions, a great effort has been devoted to the synthesis of high quality IO-Au core-shell NPs. Different methods have been reported, which are mainly classified as direct deposition and Au-seeded growth method. Different shapes of IO-Au core-shell NPs have been formed, including sphere, popcorn, star, pin, oval, and rice. This chapter describes these methods and discusses how different shapes of IO-Au core-shell NPs have been prepared. Optical properties of the resulting nanostructures are also introduced and discussed.

Pellentesque habitant morbi tristique senectus et netus et malesuada fames ac turpis egestas. Proin pharetra nonummy pede. Mauris et orci.

METHODS

One-Step Deposition Method

The simplest way to make IO-Au core-shell NPs is the one-step direct deposition method, in which Au atoms are directly deposited onto IO NPs in a growth solution via chemical reduction of the Au precursor in the growth solution. The surface of IO NPs can be different, depending on the method of preparation of the IO NPs and the growth solution. One of the most common ways to make IO NPs is coprecipitation of ferrous chloride ($FeCl_2$) and ferric chloride ($FeCl_3$). When these IO NPs are used, the gold can be coated by reduction of chloroauric acid ($HAuCl_4$) with sodium citrate at boiling temperature [22, 31, 32] or sodium borohydride ($NaBH_4$) at room temperature [33]. The thickness of Au shell can be tuned by simply varying the ratio of IO NPs and Au precursor solution. Usually, coating IO NPs with Au shell leads to decreased magnetic properties as Au is diamagnetic. However, the studies by Banerjee et al. showed that when Fe_3O_4 NPs are coated with Au, there was a distinct enhancement of magnetization by a factor of six when small Fe_3O_4 NPs (3 ~ 4 nm) were coated with a thick shell of Au (>50 nm in shell thickness) [32]. The increase of the magnetization was attributed to the formation of a large orbital magnetic moment at the interface of IO and Au.

Another common way to make IO NPs is by thermal decomposition. [34, 35]. IO NPs prepared via thermal decomposition are highly monodisperse and their sizes can be tuned with precise dimension. These nanoparticles are capped with oleic acid (OA) and/or oleylamine (OAM) and are only dispersible in organic phase. When these IO NPs are used as cores, reduction of Au precursor needs to be done in the organic phase [12, 20, 36-41]. For example, IO-Au core-shell NPs have been synthesized by mixing Au acetate ($Au(OOCCH_3)_3$), 1,2-hexadecanediol, and OA-capped Fe_3O_4 NPs in phenyl ether with heating at 180-190°C for 1.5 h [36]. The 1,2-hexadecanediol served as a mild reducing agent.

Iterative Deposition

A common way to make IO-Au core-shell NPs is to iteratively deposit Au atoms onto IO NPs by reduction of Au salt with a reducing agent [9, 10, 23, 29, 42, 43]. This repeated deposition leads to uniform Au surface while a single deposition usually makes jagged surface. The choice of reducing agent is critical. The reducing agent should promote surface catalysed reaction. A widely used surface catalysed reducing agent is hydroxylamine (NH_2OH) [44]. As demonstrated by Lyon et al. [42], uniform IO-Au core-shell NPs can be obtained after five times of addition of Au^{3+} to γ-Fe_2O_3 or Fe_3O_4 NPs prepared by coprecipitation (Figure 1A-D). In contrast, one-time reduction led to IO-Au core-shell NPs with jagged Au surface, as visualized by transmission electron microscopy (TEM) imaging. Subsequent reduction and deposition steps are needed to fill the empty sites on the IO surface without increasing the overall particle size. The core-shell NPs were confirmed by the presence of both Au and Fe peaks in the energy disperse spectrum (EDS) (Figure 1E). The IO NPs do not show distinct absorption peaks in UV-vis spectra (Figure 1F). After coating with Au, the particles give characteristic localized surface plasmon resonance (LSPR) peak in the UV-vis spectra arising from the Au shell. The LSPR peak blue shifted with the iterative deposition of Au atoms.

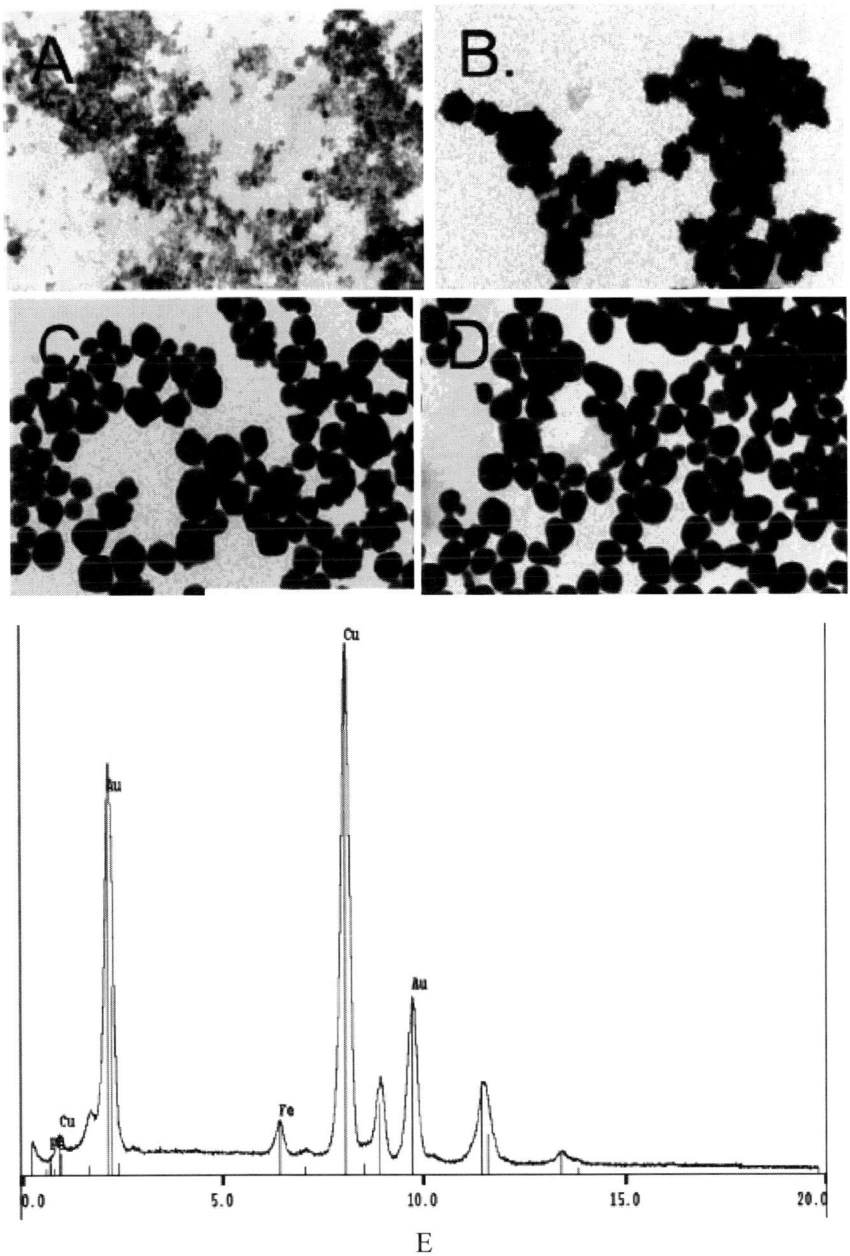

Figure 1. (Continued)

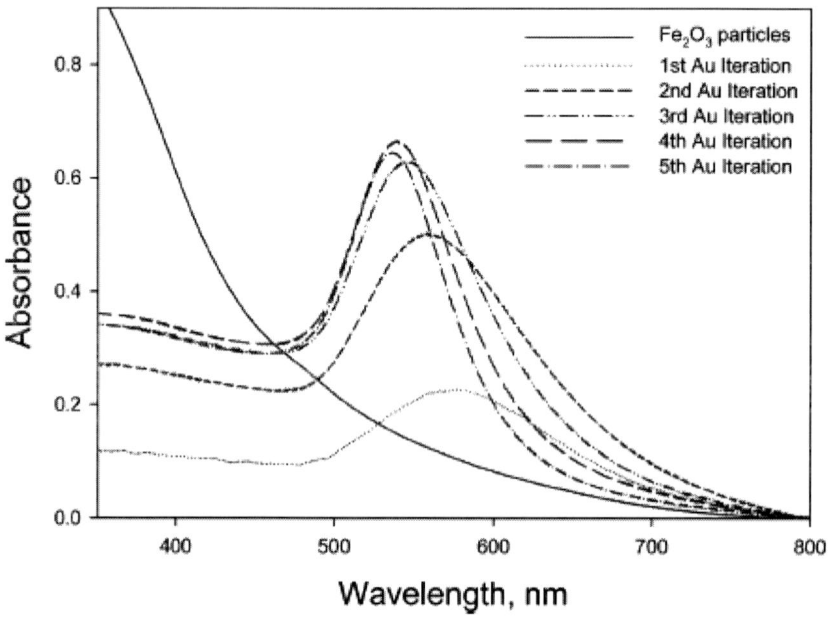

Figure 1. TEM images of IO NPs (A) and IO-Au core-shell NPs obtained with (B) one, (C) three, and (D) five times of additions of Au^{3+} to the IO NPs. (E) EDS spectrum of the sample in (C). (F) Absorption spectra of γ-Fe_2O_3 NPs and γ-Fe_2O_3 NPs after different times of addition of Au^{3+}. Reprinted with permission from ref 42. Copyright (2009) American Chemical Society.

Microemulsion Method

Earlier studies to prepare IO-Au core-shell NPs used a microemulsion (μE) [45] or nanoemulsion (nE) [46] method. The μE or nE systems can confine Au precursor and IO NPs in close proximity that facilitates the growth of Au shell. To form emulsion, a surfactant solution such as CTAB or poly (vinylpyrrolidone) (PVP) in organic solvent is firstly mixed with an aqueous mixture of $FeCl_3$, $FeCl_2$ and $HAuCl_4$. This is to form IO NPs inside the micelle via coprecipitation under a base solution such as sodium hydroxide (NaOH). To form IO-Au core-shell NPs, a reducing agent, such as sodium borohydride ($NaBH_4$), is added to reduce Au^{3+} on the surface of the IO NPs to form the Au shell.

Glue Method

To facilitate the formation of the Au layer, Jin has developed a glue method to make very small IO-Au core-shell NPs using poly-L-histidine that chelates Au^{3+} on the surface of IO NPs [7]. Specifically, they firstly modified OA-capped Fe_3O_4 NPs with amphiphilic phospholipid-polyethylene glycol to make the particles water soluble (Figure 2A). Then, the particles were coated with poly-L-histidine to immobilize Au^{3+} on the surface of IO NPs via chelation. The poly-L-histidine is capable of immobilizing Au^{3+} on IO NPs with high packing density. Subsequent reduction of the adsorbed Au^{3+} led to the formation of 1-5 nm of Au shell that is separated from the IO core with the PEG gap. The total size of the IO-Au core-shell NPs was only around 35 nm, with Au shell of 2-3 nm (Figure 2B&C). They can further vary the shell thickness by adjusting the ratio of the amount of Au^{3+} to the IO NPs, with increasing ratio leading to thicker Au shell. The NPs made by this method has near infrared (NIR) absorption with LSPR wavelength over 600 nm (Figure 2D). These particles were separated from solution when an external magnetic was attached, which demonstrated that the particles were indeed core-shell NPs not solid Au NPs (Figure 2E).

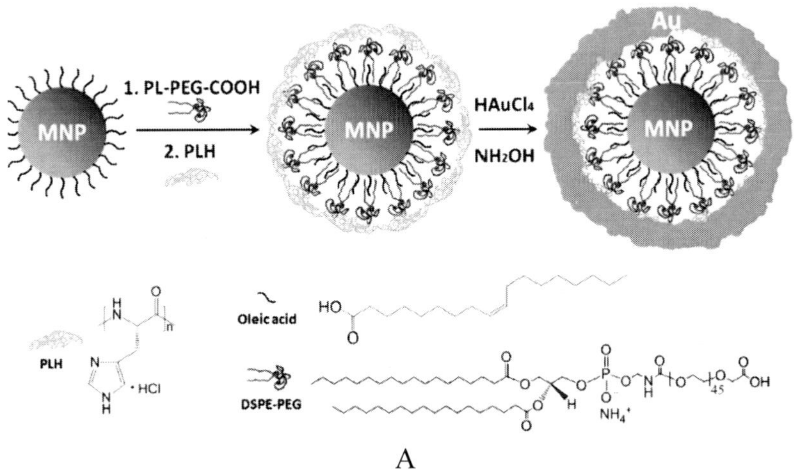

A

Figure 2. (Continued)

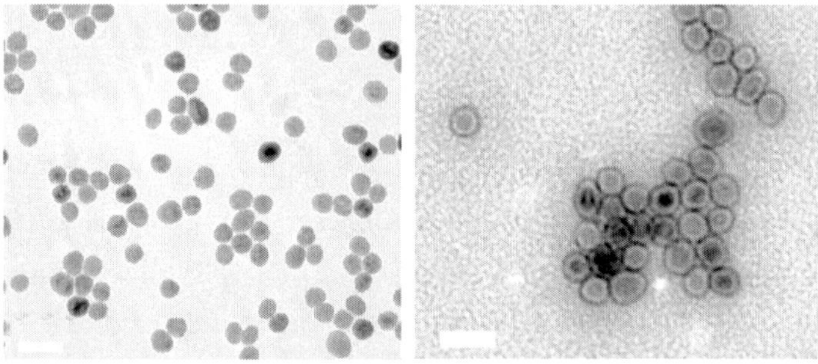

B C

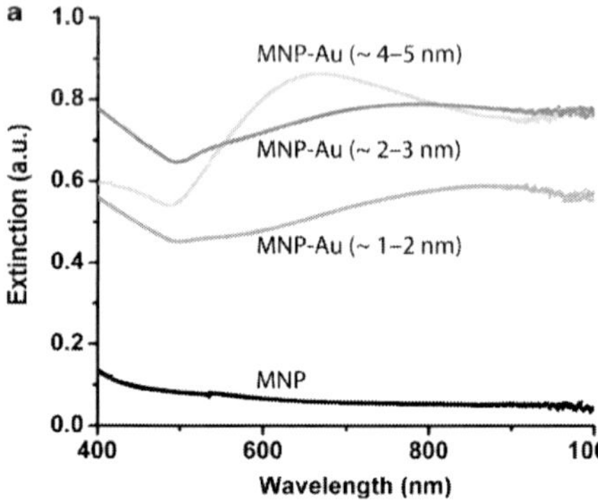

D

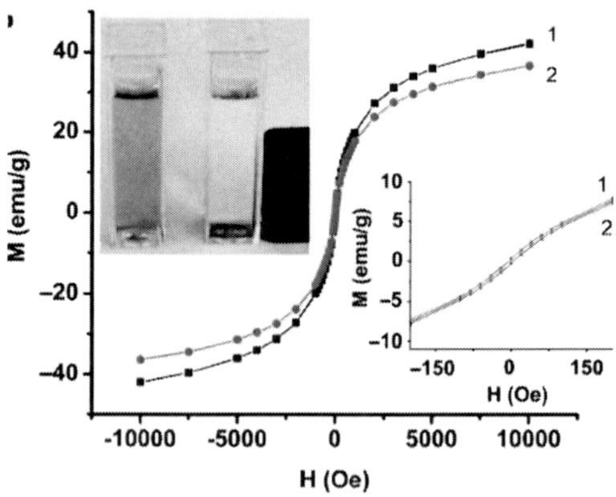

E

Figure 2. (A) Schematic of the synthesis of IO-Au cores-shell NPs via the glue method. (B) TEM image of IO NPs. (C) TEM image of IO-Au core-shell NPs. (D) Optical properties of IO-Au core-shell NPs with different shell thickness. (E) Magnetization as a function of magnetic field at room temperature for IO NPs and IO-Au core-shell NPs. Reprinted by permission from Macmillan Publishers Ltd: [*Nature Communications*] (ref 44), copyright (2010).

Seeded Growth Method

Synthesis of Spherical IO-Au Core-Shell NPs

The most widely used method to make IO-Au core-shell NPs is seeded growth method. According to typical seeded growth procedure, Au seeds are first adsorbed onto IO NPs via electrostatic interactions. The Au seeds serve as nucleation sites to initiate Au shell formation on the surface of IO NPs. Reduction of Au^{3+} in the presence of Au-seeded IO NPs leads to IO-Au core-shell NPs. The Au seeds are usually synthesized by reduction of Au^{3+} with $NaBH_4$ in sodium citrate or with tetrakis(hydroxymethyl) phosphonium chloride (THPC). These Au seeds are negatively charged due to the negatively charged capping agents. The IO NPs should thus be positively charged to attract Au seeds. One way to make positively charged IO NPs is to modify IO NPs with 3-aminopropyltrimethoxysilane (APTMS) [13, 47, 48]. In this approach, OA-capped IO NPs are first modified with OH groups using tetraethylorthosilicate (TEOS) via the sol-gel process [13] (Figure 3). APTMS was attached to the particles via silane ligand exchange. The APTMS offer positively charged amine groups to attract negatively charged Au seeds. The THPC-reduced Au seeds are adsorbed to the APTMS-modified IO NPs after overnight incubation at 4°C. To facilitate Au^{3+} reduction on IO NPs, the $HAuCl_4$ precursor was hydrolyzed to Au hydroxide with potassium carbonate (K_2CO_3). The Au-seeded IO NPs are added into the Au hydroxide solution followed by addition of formaldehyde to initiate Au shell growth.

Another way is to coat IO NPs with poly(allylamine hydrochloride) (PAH) to adsorb Au seeds as demonstrated by Wang et al. in 2008 [49]. In this approach, Fe_3O_4 NPs were firstly made negatively charged with carboxyl groups by copolymerization of acrylamide and methacrylic. Then PAH was adsorbed via electrostatic interactions. The Au seeds were synthesized by reduction of $HAuCl_4$ with aBH_4 and sodium citrate was used as the capping agent. After adsorption of the citrate-capped Au seeds onto IO NPs, NH_2OH and $HAuCl_4$ were added to initiate Au shell formation on IO NPs.

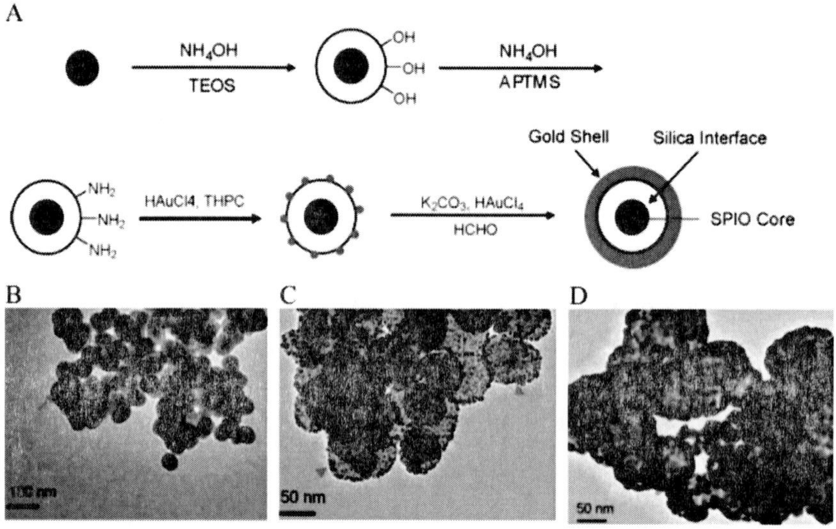

Figure 3. Synthesis of IO-Au core-shell NPs via a seed-mediated growth method. (A) Schematic of the synthesis procedure. (B) TEM image of IO NPs. (C) TEM image of Au-seeded IO NPs. (D) TEM image of IO-Au core-shell NPs. Reprinted with permission from ref 13. Copyright (2007) American Chemical Society.

Polyethyleneimine (PEI) is another positively charged polymer that is often used to adsorb negatively charged Au seeds [26, 50, 51]. An early study was done by Goon et al. in 2009 [50]. In this study, PEI-coated cubic IO NPs were directly synthesized by precipitation of iron sulfate in a basic solution, followed by oxidization with potassium nitrate at 90°C. In this solution, branched PEI (MW ~25,000) was added before reaction so that the as-prepared IO NPs were directly coated with PEI. After stirring the mixture of citrate-capped Au seeds with the PEI-capped Fe_3O_4 NPs for 2 h, Au seeds are heavily coated onto the IO NPs, with an Au density of 47.7 wt%. These seeded IO NPs were stabilized with another layer of PEI coating. The advantage of this method is the high density and high stability of the Au-seeded IO NPs that are critical to successfully make uniform IO-Au core-shell NPs.

Synthesis of Anisotropic IO-Au Core-Shell NPs

Anisotropic IO-Au core-shell NPs can be made via tuning the shape of the IO core or the Au shell. When the shape of the IO core is large and anisotropic, the NPs after Au coating will preserve the shape of the core. For example, Halas and co-workers reported the synthesis of IO-Au core-shell nanorices using IO nanorice as the core [52] (Figure 4). The hematite nanorice has lengths of 340 nm and diameter of 54 nm, with a high aspect ratio of 6.3. The nanorice was functionalized with APTMS and then coated with THPC-capped Au seeds. Reduction of Au^{3+} with formaldehyde in the presence of the Au-seeded IO nanorices led to formation of the IO-Au nanorices. The formation of the Au shell only took 5 to 10 min. Changing the ratio of the nanorices and $HAuCl_4$ can make the nanorice with different shell thickness. The hematite nanorices have combined plasmonic properties of nanorods and nanoshells in the NIR region, with increasing thickness leading to red shifted LSPR wavelength.

A different approach to make anisotropic IO-Au core-shell NPs is to manipulating the shape of the Au shell. An early example is IO-Au nanostars (NSTs) reported by Wei and co-workers using spherical Fe_3O_4 NPs as core [53, 54]. The IO (Fe_3O_4) core was only 8.4 nm. To make the IO-Au NSTs, they firslty made spherical 13 nm of IO-Au core-shell NPs in organic phase at 190°C by reduction of gold chloride with 1,2-hexadecanediol containing 8.4 nm OA- and OAM-caped IO NPs (8.4 nm). When these small spherical IO-Au core-shell NPs were introduced a growth solution containing $HAuCl_4$, cetyltrimethylammonium bromide (CTAB), silver nitrate ($AgNO_3$), and ascorbic acid (AA), IO-Au core-shell NSTs were formed. CTAB surfactant and $AgNO_3$ additive are shape-directing agents, and AA is a surface-catalysed reducing agent [55]. Due to the shape-controlled growth solution, the spherical IO-Au core-shell NPs grew into anistropic IO-Au NSTs.

We recently developed a seed mediated method that can synthesize IO-Au core-shell NPs in different shapes [51] (Figure 5). We used octahedral Fe_3O_4 NPs as the core NPs. The IO NPs were about ~ 35 nm in edge length. They were synthesized via oxidation of the precipitate of $FeSO_4$ in a base solution with KNO_3 at 90°C. In this solution, PEI (MW ~ 25,000) was added so the as-prepared IO NPs were stabilized with positively charged PEI. Then citrate-capped Au seeds were adsorbed onto the PEI-capped IO NPs and further stabilized with another layer of PEI. Formation of IO-Au core-shell NPs in different shapes were achieved by injection of the Au-seeded IO NPs in a growth solution containing $HAuCl_2$, CTAB, $AgNO_3$, and ascorbic acid (AA). When $AgNO_3$ was absent in the growth solution, the resulting IO-Au core-shell NPs were spherical. The presence of $AgNO_3$ was critical to make anisotropic particles. When the growth solution contained 60 μM $AgNO_3$, the resulting IO-Au core-shell NPs were not spherical anymore. When the AA concentration was high (1.6 mM), the IO-Au core-shell NPs had short and wide protrusions, which were called popcorns. When AA concentration was low 0.64 mM, the resulting IO-Au core-shell NPs were NSTs that had long and sharp protrusions. By adjusting the ratio of seeds and growth solution, the IO-Au nanospheres, nanopopcorns, and nanostars with different shell thickness can be obtained.

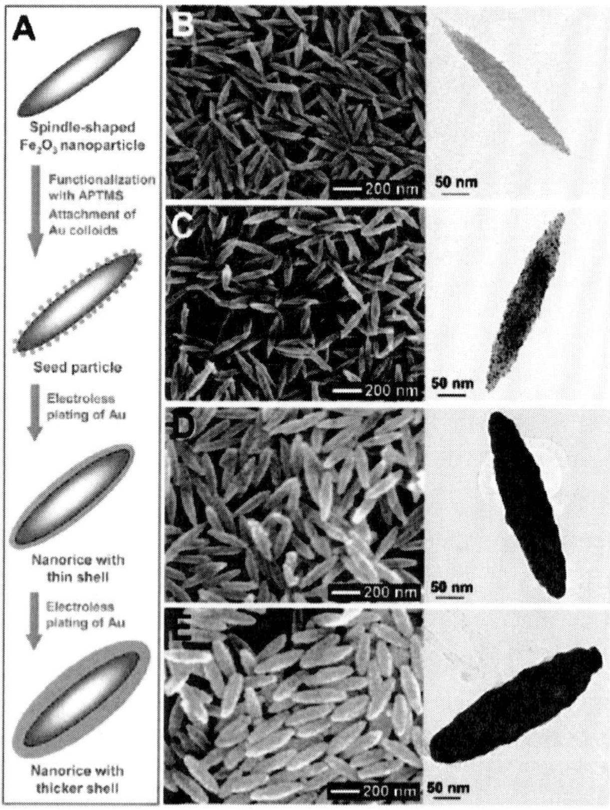

Figure 4. (Continued).

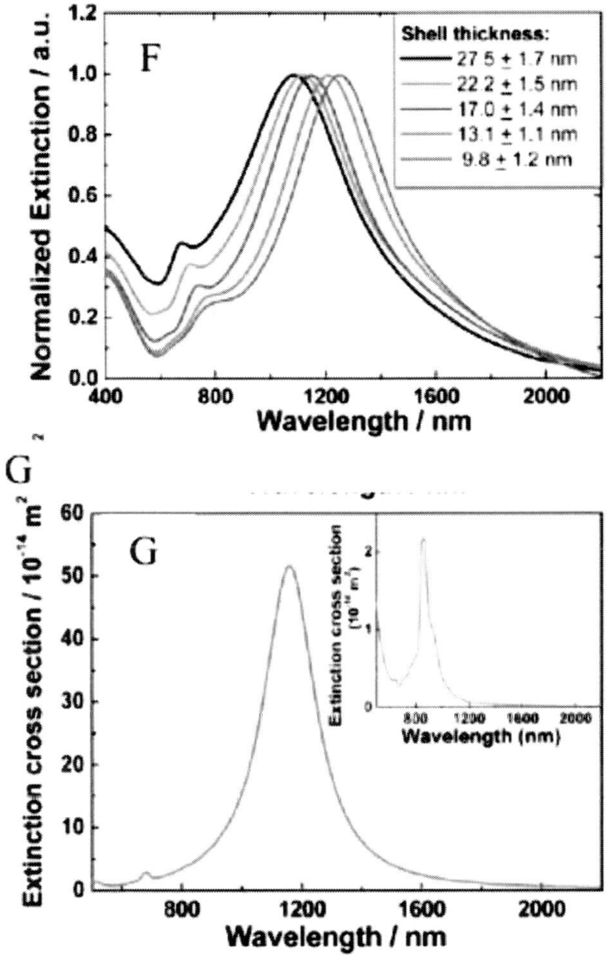

Figure 4. Synthesis and properties of IO-Au core-shell nanorices. (A) Schematic of the synthesis of IO-Au core-shell nanorices. (B-E) SEM (left) and TEM (right) images of the IO nanorices (B), Au-seeded IO nanorices (C), IO-Au core-shell nanorices with thin shells (~13 nm) (D), and IO-Au core-shell nanorices with thick shells (~28 nm) (E). (F) Absorption spectra of IO-Au nanorices with different shell thickness. (G) Calculated far-field extinction spectra of the IO-Au nanorices using FDTD. Reprinted with permission from ref 52. Copyright (2006) American Chemical Society.

The IO-Au nanospheres had LSPR in the visible region of 530-600 nm. Increasing particle's size led to a red shift of the LSPR. Compared to spheres, the popcorns had red-shifted LSPR from 600 to 670 nm depending on the size. Different from the spheres and popcorns, the NSTs gave three

LSPR peaks at 550, 700, and 850 nm. When the size of the NSTs increased, the LSPR wavelength of the peak in the visible region did not shift, but the two peaks in the NIR region red shifts. The peak in the visible region is mainly due to the plasmon resonance of the IO-Au spherical core and the two peaks in the NIR region are the results of plasmonic coupling of the core and tips. Computational studies with discrete dipole approximation (DDA) indicated that the number and length of the Au tips are the major player in the optical spectra of the anisotropic core-shell NPs. Increasing the number or the length of the Au tips red shifts the LSPR wavelength of the anisotropic IO-Au core-shell NPs.

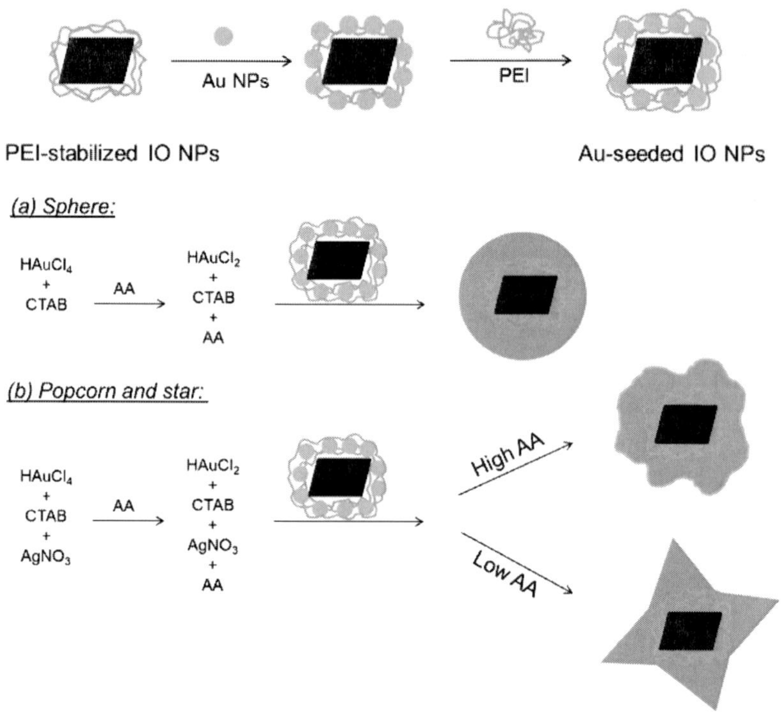

A

Figure 5. (Continued)

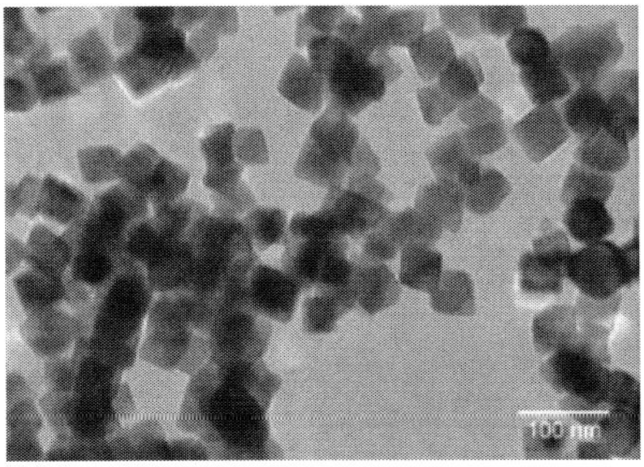

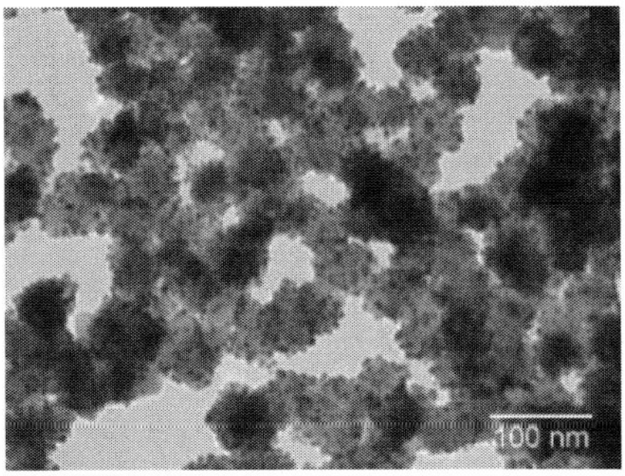

B C

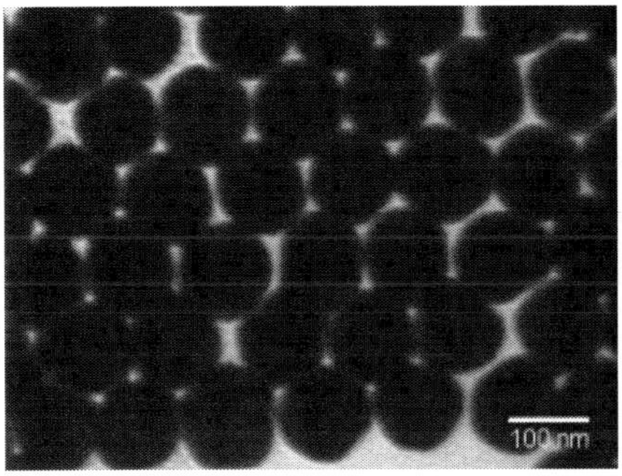

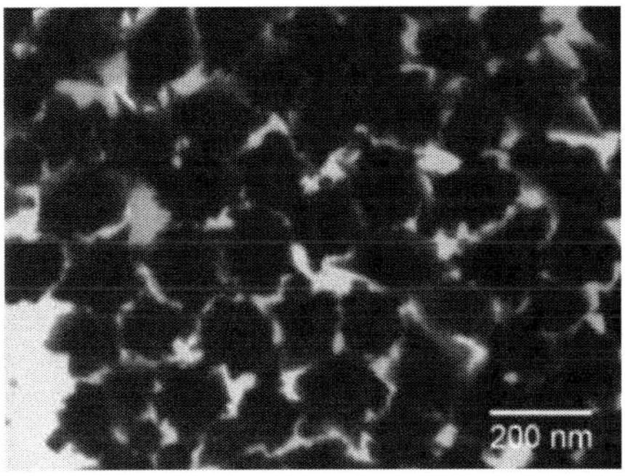

D E

F

Figure 5. (Continued)

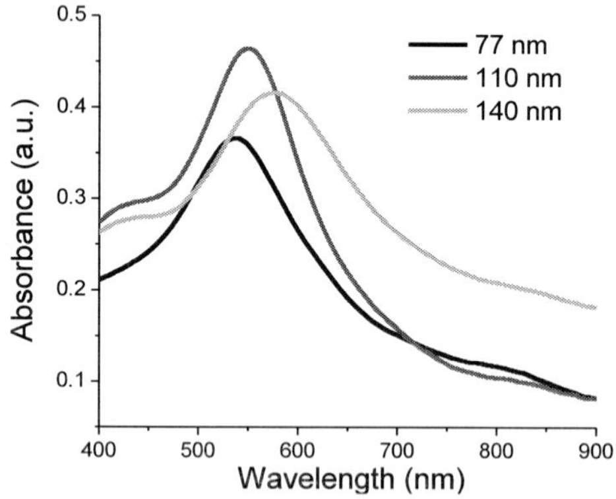

G

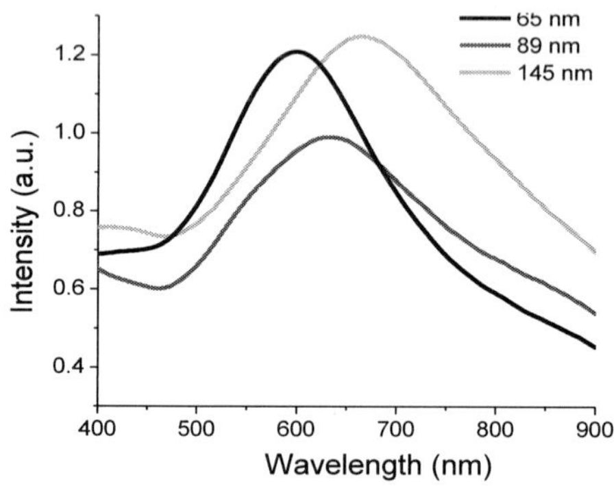

H

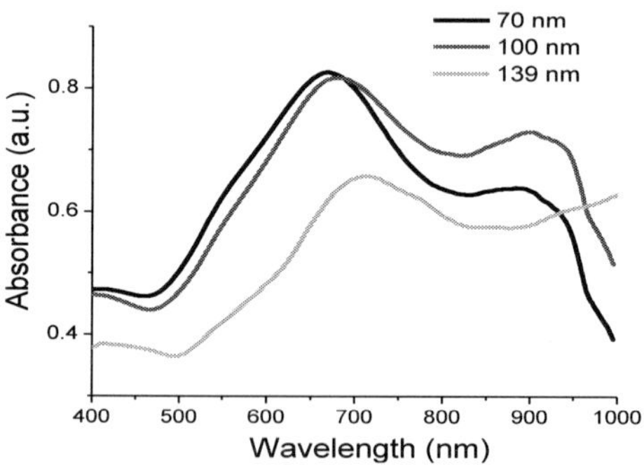

I

Figure 5. (A) Schematic of the synthesis of IO-Au core-shell NPs in spheres, popcorns, and stars. TEM image of octahedral IO NPs (B), Au-seeded IO NPs (C), IO-Au core-shell nanospheres (D), IO-Au core-shell nanopopcorns (E), and IO-Au core-shell nanostars (F). Absorption spectra of IO-Au core-shell nanospheres (G), nanopopcorns (H), and nanostars (I). Reprinted with permission from ref 51. Copyright (2016) American Chemical Society.

Anisotropic Au NPs can form because Au has different facets of (100), (110), and (111). These facets can grow in different speeds depending on the kinetics and dynamics of the growth condition [56]. As for the core-shell NPs, the adsorbed Au seeds firstly grew on the core particles and then then coalesced to form a continuous metallic shell [57]. In our recent studies, we found that the Au seeds coalesced within 5 min and formed continuous Au shells within 20 min for spheres. For popcorns and stars, no visible changes on the Au seeds were observed at 5 min [51]. Au seeds coalesced at 20 min, with continuous Au shell formation occurring after 30 min. The differences between popcorns and stars were found at 30 min, with short and wide protrusions toward popcorns and thin spikes toward stars.

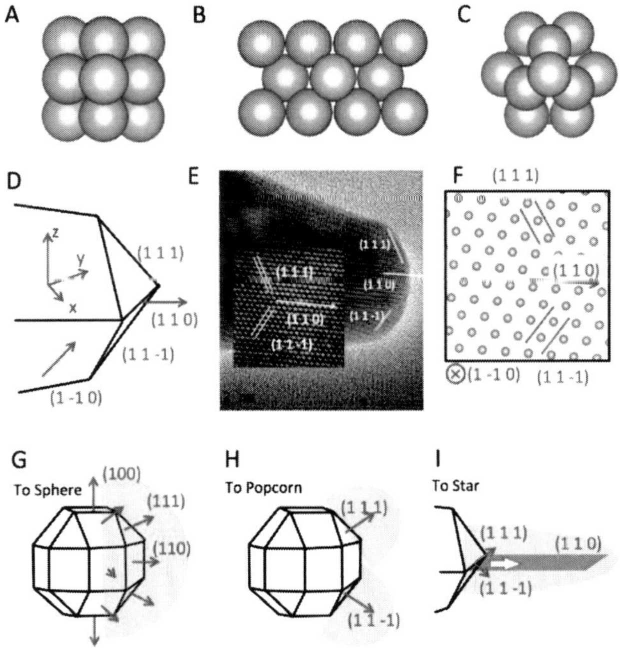

Figure 6. Computational studies on growth mechanism. Schematic of the adsorption of Ag on Au (100) surface (A), (110) surface (B), and (C) Au (111) surface. (D) Schematic and (E) HRTEM image of a tip of the nanostar. (F) Schematic of the positions of atoms based on the model in (D). (G)-(I) The proposed growth mechanism of Au nanospheres (G), nanopopcorn (H), and nanostars (I). Reprinted with permission from ref 51. Copyright (2016) American Chemical Society.

Using the principles of density functional theory method, we calculated that the adsorption energy of Ag is highest on the (110) surface, (3.0 eV/atom), followed by (100) surface (2.7 eV), and (111) surface (2.2 eV/atom). This means that the rate of Ag deposition decreases in the order of (110), (100), and (111) surfaces. Thus, in the presence of $AgNO_3$, the Au shell should grow anisotropically into difference shapes. Based on the calculation and high resolution TEM (HRTEM) images, we proposed a growth model for the formation of IO-Au core-shell nanospheres, popcorns, and stars (Figure 6). When $AgNO_3$ is absent, all the {100}, {110}, and {111} surfaces on an Au seed are available for Au deposition and thus the particles grow isotropically into a spherical shape. In the presence of $AgNO_3$, the particles grow anisotropically because of the different rates of Ag deposition on the three faces. The formation of popcorns or stars depends on the rate of Au deposition. The rate of Au deposition is determined by the concentration of AA. When the reduction of Au^+ to Au^0 is fast, the Au deposition happens at the entire (111) facet, leading to the growth of Au island along the (111) direction and thus popcorns. When the reduction of Au^+ is slow, Au atoms are mainly deposited at the ridge formed by two {111} planes. This cohesive growth led to the formation of NSTs.

Acknowledgments

We gratefully acknowledge the support from the National Institute of health (Grant No. 1R15 CA 195509-01).

References

[1] R. H. Kodama. Magnetic nanoparticles. *J. Magn. Magn. Mater*. 1999, 200, 359-372.

[2] K. J. M. Bishop, C. E. Wilmer, S. Soh, B. A. Grzybowsk. Nanoscale forces and their uses in self-assembly. *Small* 2009, 5, 1600-1630.

[3] Q. A. Pankhurst, J. Connolly, S. K. Jones, J. Dobson. Applications of magnetic nanoparticles in biomedicine. *J. Phys. D Appl. Phys*. 2003, 36(13), R167-R181.

[4] V. I. Shubayev, T. R. Pisanic II, S. Jin. Magnetic nanoparticles for theragnostics. *Adv. Drug. Deliv*. 2009, 61, 467-477.

[5] N. A. Frey, S. Peng, K. Cheng, S. Sun. Magnetic nanoparticles: synthesis, functionalization, and applications in bioimaging and magnetic energy storage. *Chem. Soc. Rev*. 2009, 38, 2532–2542.

[6] S. Singamaneni, V. N. Bliznyuk, C. Binekc, E. Y. Tsymbal. Magnetic nanoparticles: recent advances in synthesis, self-assembly and applications. *J. Mater. Chem*. 2011, 21, 16819-16845.

[7] Y. Jin, C. Jia, S. W. Huang, M. O'Donnel, X. Gao. Multifuncional nanoparticles as coupled contrast agents. *Nat. Commun*. 2010, 1(41), 1-8.

[8] H. Cai, K. Li, M. Shen, S. Wen, Y. Luo, C. Peng, G. Zhang, X. Shi. Facile assembly of Fe3O4@Au nanocomposite particles for dual mode magnetic resonance and computed tomography imaging applications. *J. Mater. Chem*. 2012, 22, 15110–15120.

[9] J. S. Aaron, J. Oh, T. A. Larson, S. Kumar, T. E. Milner, K. V. Sokolv. Increased optical contrast in imaging of epidermal growth factor receptor using magneticaly actuated hybrid gold/iron oxide nanoparticles. *Optics Express* 2006, 14(26), 12930-12943.

[10] T. A. Larson, J. Bankson, J. Aaron, K. Sokolov. Hybrid plasmonic magnetic nanoparticles as molecular specific agents for MRI/optical imaging and photothermal therapy of cancer cells. *Nanotechnol.* 2007, 18, 325101/1-8.

[11] S. J. Cho, B. R. Jarrett, A. Y. Louie, S. M. Kauzlarich. Gold-coated iron nanoparticles: a novel magnetic resonance agent for T1 and T2 weighted imaging. *Nanotechnol.* 2006, 17, 640-644.

[12] M. Kumagai, T. K. Sarma, H. Cabral, S. Kaida, M. Sekino, N. Herlambang, K. Osada, M. R. Kano, N. Nishiyama, K. Kataoka. Enhanced *In vivo* magnetic resonance imaging of tumors by

PEGylated iron-oxide–gold core–shell nanoparticles with prolonged blood circulation properties. *Macromol.* 2010, 31(17), 1521-1528.

[13] X. Ji, R. Shao, A. M. Elliott, R. J. Stafford, E. Esparza-Coss, J. A. Bankson, G. Liang, Z. P. Luo, K. Park, J. T. Markert, C. Li. Bifunctional gold nanoshells with a superparamagnetic iron oxide-silica core suitable for both MR imaging and photothermal therapy. *J. Phys. Chem. C* 2007, 111, 6245-6251.

[14] M. P. Melancon, A. Elliott, X. Ji, A. Shetty, Z. Yang, M. Tian, B. Taylor, R. J. Stafford, Chun Li. Theranostics with multifunctional magnetic gold nanoshells: photothermal therapy and T2* magnetic resonance imaging. *Invest. Radiol.* 2011, 46(2), 132–140.

[15] J. Li, Y. Hu, J. Yang, P. Wei, W. Sun, M. Shen, G. Zhang, X. Shi. Hyaluronic acid-modified Fe3O4@Au core/shell nanostars for multimodal imaging and photothermal therapy of tumors. *Biomaterials* 2015, 38, 10-21.

[16] S. Kayal, R. V. Ramanujan. Anti-cancer drug loaded iron–gold core–shell nanoparticles (Fe@Au) for magnetic drug targeting. *J. Nanosci. Nanotechnol.* 2010, 10, 1-13.

[17] R. J. Chung, H. T. Shih. Preparation of multifunctional Fe@Au core-shell nanoparticles with surface grafting as a potential treatment for magnetic hyperthermia. *Materials* 2014, 7, 653-661.

[18] Z. Fan, M. Shelton, A. K. Singh, D. Senapati, S. A. Khan, P. C. Ray. Multifunctional plasmonic shell–magnetic core nanoparticles for targeted diagnostics, isolation, and photothermal destruction of tumor cells. *ACS Nano* 2012, 6(2), 1065-1073.

[19] Y. Guo, Z. Zhang, D. H. Kim, W. Li, J. Nicolai, D. Procissi, Y. Huan, G. Han, R. A. Omary, A. C. Larson. Photothermal ablation of pancreatic cancer cells with hybrid iron-oxide core gold-shell nanoparticles. *Int. J. Nanomed.* 2013, 8, 3437-3446.

[20] C. H. Wu, Y. Y. Huang, P. Chen, K. Hoshino, H. Liu, E. P. Frenkel, J. X. J. Zhang, K. V. Sokolov. Versatile immunomagnetic nanocarrier platform for capturing cancer cells. *ACS Nano* 2013, 7(10), 8816-8823.

[21] H. Y. Park, M. J. Schadt, L. Wang, I. S. Lim, P. N. Njoki, S. H. Kim, M. Y. Jang, J. Luo, C. J. Zhong. Fabrication of Magnetic Core@Shell Fe Oxide@Au Nanoparticles for Interfacial Bioactivity and Bio-separation. *Langmuir* 2007, 23, 9050-9056.

[22] T. T. H. Pham, C. Cao, S. J. Sim. Application of citrate-stabilized gold-coated ferric oxide composite nanoparticles for biological separations. *J. Magn. Magn. Mater.* 2008, 320, 2049-2055.

[23] F. Bao, J. L. Yao, R. A. Gu. Synthesis of magnetic Fe_2O_3/Au core/shell nanoparticles for bioseparation and immunoassay based on surface-enhanced Raman spectroscopy. *Langmuir* 2009, 25, 10782-10787.

[24] U. Tamer, Y. Gu¨ndog˘du, I. H. Boyacl, K. Pekmez. Synthesis of magnetic core–shell Fe3O4–Au nanoparticle for biomolecule immobilization and detection. *J. Nanopart Res*. 2010, 12, 1187–1196.

[25] L. Lin, E. Crew, H. Yan, S. Shan, Z. Skeete, D. Mott, T. Krentsel, J. Yin, N. A. Chernova, J. Luo, M. H. Engelhard, C. Wang, Q. Li, C. J. Zhong. Bifunctional nanoparticles for SERS monitoring and magnetic intervention of assembly and enzyme cutting of DNAs. *J. Mater. Chem. B* 2013, 1, 4320-4330.

[26] L. Zhang, J. Xua, L. Mi, H. Gong, S. Jiang, Q. Yu. Multifunctional magnetic–plasmonic nanoparticles for fast concentration and sensitive detection of bacteria using SERS. *Biosens. Bioelectron.* 2012, 31, 130-136.

[27] S. Bhana, E. Chaffin, Y. Wang, S. R. Mishra, X. Huang. Capture and detection of cancer cells in whole blood with magnetic-optical nanoovals. *Nanomedicine (Lond)* 2014, 9(5), 593-606.

[28] X. Zhou, W. Xu, Y. Wang, Q. Kuang, Y. Shi, L. Zhong, Q. Zhang. Fabrication of cluster/shell Fe_3O_4/Au nanoparticles and application in protein detection via a SERS method. *J. Phys. Chem.* C 2010, 114, 19607-19623.

[29] C. H. Liang, C. C. Wang, Y. C. Lin, C. H. Chen, C. H. Wong, C. Y. Wu. Iron oxide/gold core/shell nanoparticles for ultrasensitive detection of carbohydrate–protein interactions. *Analy. Chem.* 2009, 81(18), 7750-7756.

[30] H. Chen, F. Qi, H. Zhou, S. Jia, Y. Gao, K. Koh, Y. Yin. Fe3O4@Au nanoparticles as a means of signal enhancement in surface plasmon resonance spectroscopy for thrombin detection. *Sens. Actuator B,* 2015, 212, 505-511.

[31] Q. H. Lu, K. L. Yao, D. Xi, Z. L. Liu, X. P. Luo, Q. Ning. Synthesis and characterization of composite nanoparticles comprised of gold shell and magnetic core/cores. *J. Magn. Magn. Mater*. 2006, 301, 44-49.

[32] S. Banerjee, S. O. Raja, M. Sardar, N. Gayathri, B. Ghosh, A. Dasgupta. Iron oxide nanoparticles coated with gold: Enhanced magnetic moment due to interfacial effects. *J. Appl. Phys*. 2011, 109, 123902.

[33] U. Tamer, Y. Gu¨ndog˘du, I. H. Boyacı, K. Pekmez. Synthesis of magnetic core–shell Fe3O4–Au nanoparticle for biomolecule immobilization and detection. *J. Nanopart. Res*. 2010, 12, 1187-1196.

[34] W. W. Yu, J. C. Falkner, C. T. Yavuz, V. L. Colvin. Synthesis of monodisperse iron oxide nanocrystals by thermal decomposition of iron carboxylate salts. *Chem. Commun*. 2004, 2306-2307.

[35] J. Park, K. An, Y. Hwang, J. G. Park, H. J. Noh, J. Y. Kin, J. H. Park, N. M. Hwang, T. Hyeon. Ultra-large-scale syntheses of mono-disperse nanocrystals. *Nature* 2004, 3, 891-895.

[36] L. Wang, J. Luo, M. M. Maye, Q. Fan, Q. Rendeng, M. H. Engelhard, C. Wang, Y. Lin, C. J. Zhong. Iron oxide–gold core–shell nanoparticles and thin film assembly. *J. Mater. Chem*. 2005, 15, 1821–1832.

[37] L. Y. Wang, J. Luo, Q. Fan, M. Suzuki, I. S. Suzuki, M. H. Engelhard, Y. Lin, N. Kim, J. Q. Wang, C. J. Zhong. Monodispersed core-shell Fe3O4@Au nanoparticles. *J. Phys. Chem. B* 2005, 109 (46), 21593-21601.

[38] Z. Xu, Y. Hou, S. Sun. Magnetic Core/Shell Fe3O4/Au and Fe3O4/Au/Ag nanoparticles with tunable plasmonic properties. *J. Am. Chem. Soc*. 2007, 129(28), 8698-8699.

[39] S. Pal, M. Morales, P. Mukherjee, H. Srikanth. Synthesis and magnetic properties of gold coated iron oxide nanoparticles. *J. Appl. Phys.* 2009, 105, 07B504.

[40] I. Robinson, L. D. Tung, S. Maenosono, C. Walti, N. T. K. Thanh. Synthesis of core-shell gold coated magnetic nanoparticles and their interaction with thiolated DNA. *Nanoscale* 2010, 2, 2624-2630.

[41] E. Skoropata, R. D. Desautels, C. C. Chi, H. Ouyang, J. W. Freeland, J. van Lierop. Magnetism of iron oxide based core-shell nanoparticles from interface mixing with enhanced spin-orbit coupling. *Phys. Rev. B* 2014, 89, 024410.

[42] J. L. Lyon et al. Synthesis of Fe Oxide Core/Au Shell Nanoparticles by Iterative Hydroxylamine Seeding. *Nano Lett.* 2004, 4(4), 719-723.

[43] G. V. P. Kumar, N. Rangarajan, B. Sonia, P. Deepika, N. Rohman, C. Narayana. Metal-coated magnetic nanoparticles for surface enhanced Raman scattering studies. *Bull. Mater. Sci.* 2011, 34(2), 207-216.

[44] K. R. Brown, M. J. Natan. Hydroxylamine seeding of colloidal Au nanoparticles in solution and on surfaces. *Langmuir* 1998, 14, 726-728.

[45] M. Mikhaylova, D. K. Kim, N. Bobrysheva, M. Osmolowsky, V. Semenov, T. Tsakalakos, M. Muhammed. Superparamagnetism of magnetite nanoparticles: dependence on surface modification. *Langmuir* 2004, 20, 2472-2477.

[46] H. Liu, P. Hou, W. X. Zhang, J. H. Wu. Synthesis of monosized core–shell Fe3O4/Au multifunctional nanoparticles by PVP-assisted nanoemulsion process. *Colloids Surf. A* 2010, 356(1-3), 21–27.

[47] M. R. Rasch, K. V. Sokolov, B. A. Korgel. Limitations on the optical tunability of small diameter gold nanoshells. *Langmuir* 2009, 25(19), 11777-11785.

[48] C. S. Levin, C. Hofmann, T. A. Ali, A. T. Kelly, E. Morosan, P. Nordlander, K. H. Whitmire, N. J. Halas. Magnetic-plasmonic core-shell nanoparticles. *ACS Nano* 2009, 3(6), 1379-1388.

[49] L. Wang, J. Bai, Y. Li, Y. Huang. Multifunctional nanoparticles displaying magnetization and near-IR absorption. *Angew. Chem. Int. Ed.* 2008, 47, 2439 –2442.

[50] I. Y. Goon, L. M. H. Lai, M. Lim, P. Munroe, J. J. Gooding, R. Amal. Fabrication and dispersion of gold-shell-protected magnetite nanoparticles: systematic control using polyethyleneimine. *Chem. Mater*. 2009, 21(4), 673–681.

[51] E. A. Kwizera, E. Chaffin, X. Shen, J. Chen, Z. Gai, S. Bhana, R. O'Connor, L. Wang, H. Adhikari, S. Mishra, Y. Wang, X. Huang. Size- and shape-controlled synthesis and properties of magnetic-plasmonic core-shell nanoparticles. *J. Phys. Chem*. C 2016, 120, 10530−10546.

[52] H. Wang, D. W. Brandl, F. Le, P. Nordlander, N. J. Halas. Nanorice: a hybrid plasmonic nanostructure. *Nano Lett*. 2006, 6(4), 827-832.

[53] Q. Wei, H. M. Song, A. P. Leonov, J. A. Hale, D. Oh, Q. K. Ong, K. Ritchie, A. Wei. Gyromagnetic imaging: dynamic optical contrast using gold nanostars with magnetic cores. *J. Am. Chem. Soc*. 2009, 131, 9728-9734.

[54] H. M. Song, Q. Wei, Q. K. Ong, A. Wei. Plasmon-resonant nanoparticles and nanostars with magnetic cores: synthesis and magnetomotive imaging. *ACS Nano* 2010, 4, 5163--5173.

[55] C. J. Murphy, T. K. Sau, A. M. Gole, C. J. Orendorff, J. Gao, L. Gou, S. E. Hunyadi, T. Li. Anisotropic metal nanoparticles: synthesis, assembly, and optical applications. *J. Phys. Chem. B* 2005, 109(29), 13857-13870.

[56] T. K. Sau, C. J. Murphy. Room temperature, high-yield synthesis of multiple shapes of gold nanoparticles in aqueous solution. *J. Am. Chem. Soc*. 2004, 126, 8648-8649.

[57] S. J. Oldenburg, R. D. Averitt, S. L. Westcott, N. J. Halas. Nano-engineering of optical resonances. *Chem. Phys. Lett*. 1998, 288, 243-247.

In: Nanotechnology
Editor: Paresh Chandra Ray
ISBN: 978-1-53613-889-4

Chapter 2

MULTITUDINOUS HETEROATOM DOPED CARBON DOTS: A VERSATILE NANOSCOPIC MOLECULAR VEHICLE FOR TARGETED ANTICANCER DRUG DELIVERY

Avijit Pramanik, Kaelin Gates, Ye Gao, Salma Begum, Aruna Vangara and Paresh Chandra Ray*
Department of Chemistry and Biochemistry,
Jackson State University, Jackson, Mississippi, US

ABSTRACT

Luminescent and photo responsive water-soluble, highly bio-compatible carbon dots have shown substantial healing efficiency and promising potentials in biomedical applications that include bio-imaging, multifunctional drug delivery, disease detection, integrated targeting, tumor rehabilitations and more. These nanodot embedded drug delivery

* Corresponding Author Email: paresh.c.ray@jsums.edu.

systems have been used as therapeutic agents to control drug release and deliver drugs under physiological milieu in an efficient way. In this review, we summarize the recent research in the field of targeted nano drug delivery systems using multitudinous heteroatom doped luminescent carbon dots (C-dots), functionalized or surface passivated carbon dots, nanohybrid or doped carbon dots as some delivery systems for doxorubicin (DOX), an anticancer drug. By varying the heteroatom-doped carbon materials with different carbon nano-dots nanostructures and enhancing the efficiency of DOX, the DOX loaded carbon dots could be an effective candidate to generate smart drug products and improve clinical outcome. We addressed synthetic strategies of various carbon dots, most common characterization methods, mechanism for intracellular uptake, pH-triggered drug release and monitoring clearance. Consequently, the trends of these methodologies on drug delivery, is highlighted in the breakthrough on improving the analytical performance and the emerging applications of carbon-based nanomaterials to provide recent perspective on future research. We also outlined the several unique features of drug loaded C-dots which has been used to design a nanocarrier vehicle for studying the diverse drug delivery applications.

Keywords: heteroatom doped carbon dots, doxurubicin (DOX), drug delivery, cell imaging, targeted cancer treatment

INTRODUCTION

Multifunctional metallic and nometallic nanodots are the emerging nanomaterials with numerous applications in the field of sensor technology, [1-5] photocatalysis, [6-9] light emitting device, [10-12] pharmaceutical science especially clinical diagnostics and medical research [13-21] and so on. These smart nanomaterials exhibit extraordinary optoelectronic properties, 22-24 cell imaging agents, [25-31] high drug delivery efficiency 32-38 and especially selective transports of molecular drugs to the targeted organs that could provide visual evidence for cell health and disease diagnosis. Among the recently developed nano systems, carbon nanodots represent a new class of promising alternative nanoplatform in the nanocarbon family, due to their benign nature for favorable light-driven energy conversion, exceptional biocompatibility, bio-degradability with limited life span, fine water-dispersibility, ease of surface functionalization, resistance against photo bleaching, simultaneous excitation of multiple fluorescence colors, good cell permeability and excellent signal brightness. [39-42] Carbon dots (CDs or C-dots) or carbon nanodots (CNDs), usually comprise of discrete, quasi-spherical tiny carbon nanoparticles with sizes below 10 nm, which was first isolated and discovered in 2004 from single-walled carbon nanotubes, during the purification process through preparative electrophoresis. [43] Again in 2006, carbon nanodots were also obtained from graphite powder and cement via the laser ablation method by Y. P. Sun and his group. [44] Since its serendipitous inspection, a considerable amount of work has already been done and there is still a need to explore more due to the aforementioned specific benefits of carbon dots including efficient synthetic routes that are safe, cost-effective, easily isolated and environmentally friendly. [45] Usually carbon nanodots are assembled as a different degree of (mostly sp2) hybridized carbogenic core with graphitic layers of turbostratic carbons [46] and composed with multiple oxygen-containing species (with content of 5-50 wt%) of various functional groups represented by aldehyde, epoxy, ether, carbonyl, hydroxyl, and carboxylic acid on their surface. [47] These terminated oxygenic moieties in carbon dots provide potential for further functionalization and better crystallinity than its cousins, and imparts higher dispersity over a wide range of solvents, including water, than other nanomaterials, such as graphene quantum dots, semiconductor quantum dots, metal nanodots etc. On the other hand, due to quantum confinement effect of conjugated-domains, intrinsic band gap, homogeneous surface defects with zigzag edges and electron donating and accepting nature in a single system, the carbon nanodots (C-dots) are endowed with distinct properties. These properties include effective absorbance of sunlight, tunable photoluminescence visible to near-infrared region, photobioluminescence, [48] electrochemiluminescence [49] and more, with wide potentials in the field of chemical science and nanotechnology. The unique optoelectronic properties of C-dots can be tuned by the doping of heteroatoms such as nitrogen, [50] phosphorous, [51] boron [52] or sulfur [53] etc., which leads to an enhanced quantum yield with higher photostability, ease of surface functionalization, as well as efficient sensitivity and selectivity. These heteroatom doped carbon dots have the potential to pursue nontoxic alternatives of currently used heavy metal-based semiconductor quantum dots (Scheme 1).

Designing medically useful nanodevice for localized drug delivery with specific targeting, disease prevention and health promotion, are still the major challenges for modern medical technology. To develop modern nanomedicines as well as nanodevices than can function inside the body against dreaded diseases, it is necessary to understand on both precise biochemical fundamentals and observable biochemical activity reporters. Chemotherapy is the currently only efficient widely known technique, to destroy dividing cancer cells by using many drugs followed by several cycles for number of times. The therapeutic index in drug development can be improved by delivery or uptake of the drugs in target organs through the appropriate use of nanotechnologies. Recently, the Food and Drug Administration (FDA) also approved many targeted cancer therapies to treat specific types of cancers, and few are in preclinical trial with animal models. However, these types of therapeutic techniques are still non-specific which destructively impact both cancerous and healthy cells. The significant challenge in cancer chemotherapy is to understand the mechanisms of most complex diseases by the drug distribution and release to the targeted cancer cells in a controlled way to the organs. Therefore, it is critically important to develop some sophisticated strategies to achieve both targeted and traceable anti-cancer drug delivery. Among the newly developed nanomedicine and nanodevices, the carbon based nanomaterials possess several advantages including exceptional biocompatibility, hydrophilicity and more, which can be good candidates for cell imaging and sensor technology in clinical and translational research. [54, 55] In particular, organophilic/hydrophobic multifunctional C-dots with large surface to mass ratio confers the ability to absorb and carry many drug molecules that facilitate fast penetration of cell membranes and would be exploited as a versatile drug delivery vehicle for chemotherapeutic payloads. In addition, C-dots also act as biocompatible fluorescent nanocarriers in the construction of a photoresponsive drug delivery system with both fluorescence and photo-controlled release functionalities. The functionalities facilitate fast penetration of drug loaded nanosystems to the cell membrane and the organelles to interact directly with DNA and proteins, which ensure high drug delivery efficiency as well as drug release. Moreover, aromatic drugs can be loaded with sp2 and sp3 hybridized multifunctional C-dots or doped C-dots via noncovalent electrostatic interactions, making them promising drug delivery systems for cancer treatments. The drug loaded nanocarriers a meant to improve efficacy of cancer drugs by potentially lowering the doses (concentrations), enabling different precision medicines, specific targeting drug delivery, improving the timed release of drug molecules and reducing the toxicity of a cancer drug during therapeutic action, leading to the fast development of new safe and biocompatible medicines.

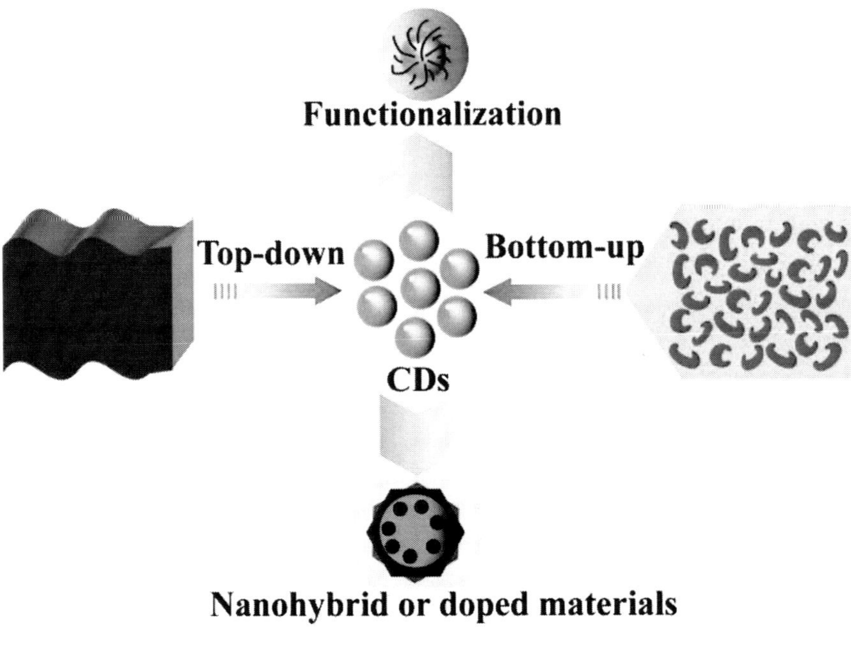

Scheme 1. Schematic synthetic strategies for carbon quantum dots (CDs) or Carbon dots (C-dots) via "top-down" and "bottom-up" methods, and modification including passivation, surface passivation and functionalization or doping and nanohybrids.

Cellular Uptake Mechanism and Doxorubin-Loaded Environmental Friendly Doped Carbon Dots for Targeted Anticancer Drug Delivery

In this respect, multifunctional C-dots or doped C-dots are loaded with doxorubicin (DOX), a selective anticancer drug, for nanomaterial-based extracellular microenvironment triggered targeting anticancer drug delivery systems where multicolor nanodots are used as effective trafficking agents for drug nanocarriers. All these anticancer drugs are used for treating many cancers, such as breast cancer, lung cancer, lymphoma, ovarian cancer, and other malignant tumors. These drugs are intercalated with DNA, breaking the DNA chain and blocking the activity of topoisomerase II, thereby stopping the process of replication. Currently, Doxorubicin (DOX) (Figure 1) is one of the most leading targeted anticancer drugs with broad spectrum activity. It belongs to the Anthracyclines family which is basically used for cancer chemotherapy and is extracted from Streptomyces bacterium such as Streptomyces peucetius. DOX exhibits potent anticancer action, due to the DNA damage through topoisomerase II inhibition. DOX also generates free radicals, which induces oxidative stress followed by apoptosis. However, the exact molecular mechanism is still unclear. Free radical generator of DOX induced by the oxidative stress increasingly causes low-therapeutic index, cardiac toxicity, bone marrow suppression, rapid drug resistance, genotoxicity and other side effects in normal human cells and is potentially carcinogenic in animal as well. In an effort to reduce the adverse side effects of DOX, researchers introduced selective nanocarriers that can increase the bioavailability of a drug, can be ameliorated to promote the targeted therapeutic potency with reduced side effects and enhance the treatment effects. Cellular uptake and intracellular trafficking of multifunctional C-dots or drug loaded C-dots by eukaryotic cells may occur through endocytic pathways which includes phagocytosis, and pinocytosis mechanistic routes. [56, 57] In pinocytosis route is described as the uptake of fluids and molecules within small vesicles whereas phagocytosis involves the engulfing of comparatively bigger molecules especially microorganisms and cell debris. However, many factors such as particle size, shape, the specific surface charge, surface area, surface curvature affect, cell type, and even culture condition, are involved in the selection of the cellular uptake pathways. Even the uptake efficiency might even be affected by specific internalization of cells and C-dots interactions. This can be modulated by changing cell-specific parameters such as cell type or cell cycle phase. On the other hand, uptake efficacy can be monitored by changing the charge density of the cell membrane and C-dots complexes. The cell membrane consists of a lipid bilayer with an anionic surface that can be neutralized by the uptaking of cationic C-dots. Once the lipid membrane and C-dots complex fall to neutral, the charge density as well as interactions become weakened which will also increase the aggregation of complexes, becoming bigger particles and hard to be internalized the C-dots into cells. As a result, uptake efficiency also inhibited a lot. Although, recent report says that uptake efficiency of protein-coated spherical gold nanoparticles is more efficient than rod-shaped gold nanoparticles with three different cell lines. [58, 59] Koping-Hoggard et al. [60] recently found that due to aggregation, the oligomeric chitosan can form multiple number of heterogeneous nanostructures, that can be tuned by changing the sizes and shapes and other experimental conditions. In this regard, the globular nanostructures also increase with increasing chain length, increasing the charge ratio as well as the reduction of pH (from 6.5 to 3.5). However, the uptake efficiency mechanism is still very complicated and leaves more space for researchers to discuss regarding these issues. As shown in scheme 2 by Wang et al., [61] clearly demonstrated the cellular uptake and intracellular trafficking mechanisms with DOX loaded C-dots, where DOX loaded C-dots are internalized via membrane penetration into the cells by endocytosis. DOX loaded C-dots can also be recycled back to the plasma membrane in this mechanism. Upon endocytosis, DOX loaded C-dots are enclosed within the early endocytic vesicles and then the DOX loaded C-dots vesicles were transported into the lysosomes for degradation or recycle back to the Golgi. The protonated DOX was free from C-dots due to the acidic environment.

Finally, the DOX was released and entered the nuclei. For this study, this group synthesized ultra-small hollow C-dots by adopting freeze–dried BSA (Bovine serum albumin) protein as a carbon source by the method of hydrothermal treatment followed by centrifugation and dialysis. Before solvothermal treatment, the BSA solution was dispersed in water under sonication. In this process, BSA monomers tend to aggregate overtime even at high concentration with uniform sizes. During this process, BSA was denaturized with non-uniform sizes ranging from ca. 10 nm to 80 nm. However, after a long hydrothermal treatment, hollow C-dots were obtained. Since BSA contains several amino acid groups, the resulting C-dots are also doped by many heteroatoms including oxygen, nitrogen and sulfur. Due to incomplete carbonization, the surface of the nanoparticles may prevent the shell from sinking or breaking, which was confirmed by STEM mapping analysis. Attachment of DOX on hollow C-dots was carried out simply by mixing an aqueous solution of DOX (0.1mg/mL) to an aqueous solution of hollow C-dots (0.1mg/mL) under dark conditions and stirring overnight at 4°C. Then the mixture was dialyzed against DI water to remove unbound DOX completely and insoluble impurities were then removed by filtration. The accumulation of DOX into hollow C-dots (loading ratio) can be determined based on the comparative absorbance, photoluminescence spectra, and surface zeta potential measurements of DOX and hollow C-dots. It is also noted that the water dispersed DOX loaded hollow C-dots were very stable without any noticeable precipitation for several weeks and displayed red fluorescence under UV irradiation. In order to study the controlled release of drugs from the carrier, DOX loaded hollow C-dots were transferred to the fresh dialysis bag and dialyzed against phosphate buffered saline (PBS) at pH of 7.4 (physiological pH) and 5.0 (acidic pH). This was done at 37°C and was kept in dark phase under slow stirring to allow spontaneous release of DOX from release system. In this aspect, DOX loading efficiency (% DLE) and percentage of cumulative drug release (%CR) could be calculated according to following equations:

$$\%\ \text{cumulative drug release} = \frac{\text{Initial conc. of DOX } (C_0) - \text{Final conc } (C_f)}{\text{Initial conc. of DOX } (C_0)} \times 100$$

$$\%\ \text{of drug loading efficiency} = \frac{\text{Theoretical amount of drug loaded } (A_{DOX}) - \text{Free drug } (A_{Free})}{\text{Theoretical amount of drug loaded } (A_{DOX})} \times 100$$

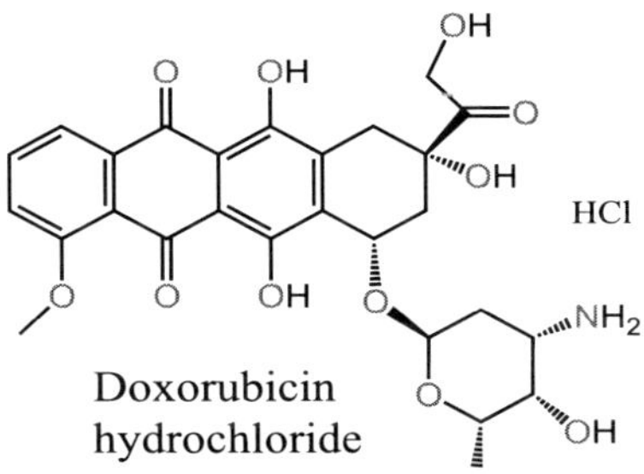

Figure 1. Chemical structure of anticancer drug doxorubicin hydrochloride (DOX).

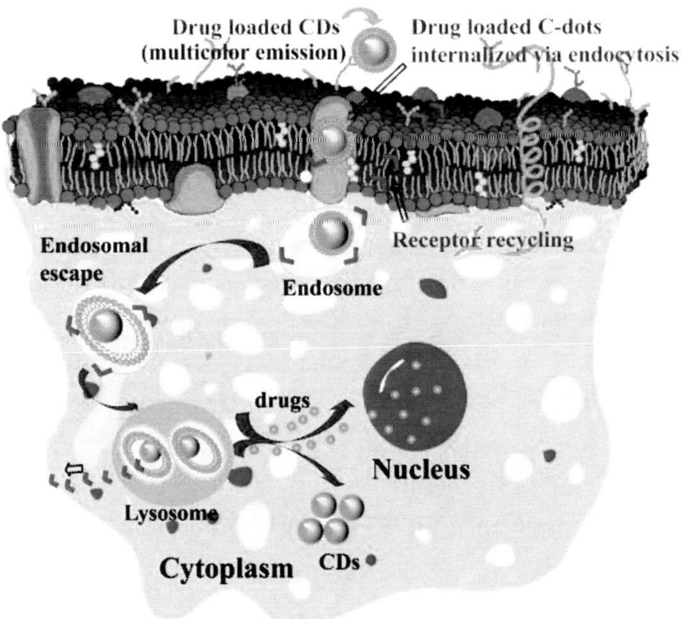

Scheme 2. Schematic illustration of the intracellular release of anticancer drugs form the anticancer drug loaded-CDs drug delivery systems.

To study the amount of DOX release, DOX concentration was quantified by collecting the sample from the dialysis bags at different time intervals using spectroscopic analysis. *In vitro* studies demonstrated that the release of DOX under acidic environment at pH 5.0, which mimics tumor microenvironment or intracellular lysosomes, was significantly faster (~70%) compared to physiological pH (~4%) that mimics extra-cellular environment. These drug release studies illustrated that DOX solubility is higher in water under acidic conditions. To investigate cellular uptake, A549 cells were used and treated with DOX loaded hollow C-dots or free DOX for different total incubation periods. Fluorescence microscopy images confirmed that intense red fluorescence emission from DOX loaded C-dots can be observed in the cells but no obvious red fluorescence from free DOX was seen in the cells after incubation for few hours. In addition, the DOX loaded C-dots entered the cells faster than free DOX. This is due to the DOX loaded hollow C-dots following the endocytosis pathway for cellular uptake, however, free DOX may uptake by cells via flip-flop mechanism. [62] To investigate the *In vitro* cytotoxicity of hollow C-dots and DOX loaded hollow C-dots, A549 cells and another four different cell lines were used and incubated with wide concentration ranges of hollow C-dots and DOX loaded hollow C-dots with different exposure time from 24 to 48 h. The cell viability and cytotoxicity assay results show that no growth inhibition could be detected after incubation for 48 h, even at 400 mg/mL hollow C-dots. However, maximum 100 mg/mL concertation was used for drug delivery experiments. The cytotoxicity of free DOX is significantly higher than DOX loaded hollow C-dots, which were determined by CCK-8 assay. However, several reports highlighted that few carbon based nanomaterials sometime show more than 100% cell viabilities than control, [63] may stimulate cell growth, promote cells proliferation, scattering, differentiation and migration by structural scaffolds. [64] Ding et al. [65] presented a novel DNA-C-dots that was synthesized from DNA, serve as a fluorescent vehicle for anticancer drug delivery (Scheme 3a). In brief, DNA was extracted and isolated from Escherichia coli in lysogeny broth media for 24 h at 37°C. [66] Then the purified DNA was obtained by high speed centrifugation and the solid pellet was resuspended in water. The DNA-C-dots have been synthesized by the hydrothermal method followed by centrifugation to obtain solid and then purified using PCR purification kit followed by complete dialysis for 3 days. In contrast to other common reported carbon dots, synthesized using facile bottom-up methods, including hydrothermal or solvothermal carbonization, acid hydrolysis, and microwave, the DNA-C-dots are practically reluctant for dehydration or oxidation (Scheme 3b). Structural analysis reveals that the DNA and the DNA–CDs, both are stabilized by mostly common covalent bonds, which are also confirmed by XPS, ^{1}H-NMR, ^{13}C-NMR, ^{31}P-NMR and FT-IR spectroscopy and elemental analysis. N1s XPS data also revealed that as-synthesized DNA–CDs exhibit new bonds (N–O and N–P), compared to natural DNA. However, elemental analysis reveals that DNA containing N elements are significantly decreased in DNA–CDs. This may be due to certain NH_2 groups possibly converting into N_2 gas that was released during the hydrothermal treatment. In addition, the formation of N–O and N–P bonds through the hydrothermal carbonization of the DNA are due to the consequences of dehydration, condensation, and carbonization reactions between the nitrogenous bases of DNA: adenine, thymine, cytosine, and guanine and five-deoxyribose sugars interacting with neighboring phosphate groups (phosphorus and oxygen monomer). For the preparation of drug loaded DNA-C-dots complexes, the DNA-C-dots solution and doxorubicin solution (2:1 ratio, w/w) were mixed with PBS buffer at pH 7.4 under stirring for a few hours and then allowed to react for overnight in an incubator shaker with 200 rpm speed at 24°C. The resulting solution was dialyzed against deionized water for 2 days using a dialysis membrane (MWCO-3500 D) to remove any unreacted DOX. This group also used another set of molecules (Rh 6G and propidium iodide) for drug loading and drug release studies. The pH-dependent *In vitro* drug release tests were conducted on a shaking table in dark conditions. In these experiments, aliquots of DNA-C-dots-DOX or DNA-C-dots-Rh 6G were poured in a dialysis bags (MWCO-1000 Da) and dialyzed against buffer solution across the range of different pHs (1X PBS buffer at pH 7.2 and 20 mM HAc-NaAc buffer at pH at 4.1) to simulate a similar microenvironment that closely mimics human tissue such as physiological environments, interstitial fluid of cancer and endochylema of cancer cells. [67] Then at the selected time intervals, aliquots of the same amount from outside of the dialysis bags were taken out to quantify the released drugs from DNA-C-dots-DOX or DNA-C-dots-Rh-6G complexes by fluorescence spectrophotometry as described previously. To validate the potential applications, especially live cells imaging and drug delivery, the microbes, such as E. coli and S. cerevisiae cells were used to evaluate their internalization capability toward DNA-C-dots. Both the cells were separately being incubated with DNA-C-dots in Lysogeny broth (LB) or yeast extract peptone dextrose (YPD) media for 6 h at 37 °C. Then all cell pellets were carefully washed 3 times with PBS buffer (pH 7.4) prior to imaging analysis by confocal fluorescence microscopy (CFM). Notably, Cytotoxicity tests of DNA-C-dots for both the cell lines exhibited almost zero toxicity and high biocompatibility, which means that DNA-C-dots can be used in either bacterial or eukaryotic cells studies. Fluorescence microscopy images demonstrated that DNA-C-dots can readily enter either E. coli or S. cerevisiae cells and emit bright green fluorescence under the excitation of 405 nm laser. In order to examine *In vivo* drug loading efficiency and drug release behavior by DNA-C-dots vehicles, rhodamine 6G (Rh 6G) and doxorubicin (Dox) were administered directly without a carrier as test molecules loaded into DNA-C-dots vehicles. Due to electrostatic interactions supported by zeta potentials data, the positively charged vehicle (DOX or Rh-6G) and negatively charged cargo (DNA-C-dots) may have interacted with each other and formed a stable supramolecular complex at a neutral pH. Although, a lower pH may trigger drug release from the DNA-C-dots−Rh 6G (DNA-C-dots−Dox) nano-complex systems due to higher solubility at mild acidic condition. The intracellular optimal uptake of DNA-C-dots −Rh 6G or DNA-C-dots−DOX nanocomplex into the S. cerevisiae cells and their fluorescence imaging, transport and delivery of Rh 6G or DOX by DNA-C-dots, were shown in Figure 1a−f. The DOX was slowly internalized by the cells and released DOX molecules to the cells to keep a balance of adsorption and desorption through electrostatic interactions.

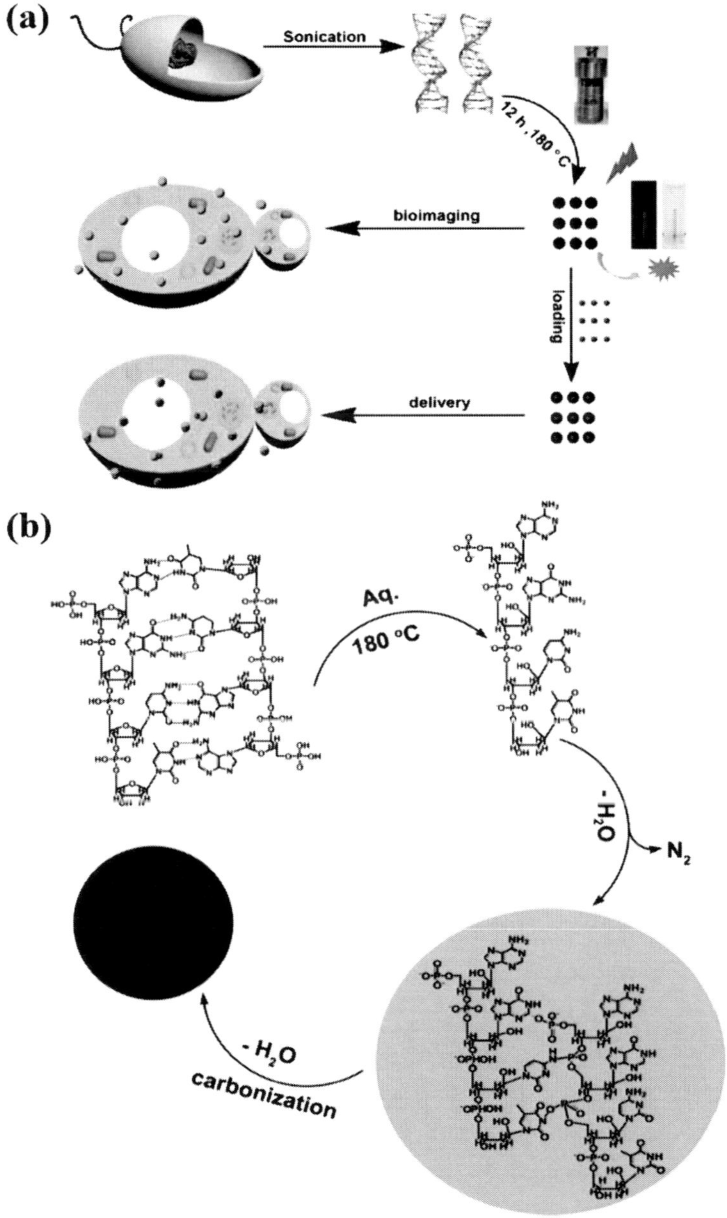

Scheme 3. (a) Schematic Representation of DNA-C-dots Synthesis and Its Application in Bioimaging and Drug Delivery, (b) Proposed mechanism for DNA-C-dots synthesis: denaturation, condensation, polymerization, and carbonization (Reproduced with permission from reference, [65] (Copyright 2015 American Chemical Society).

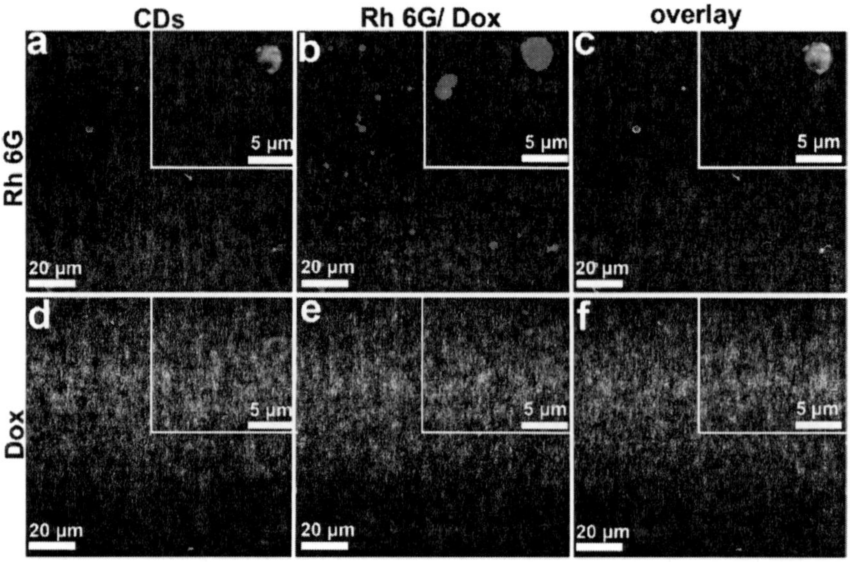

Figure 2. DNA-C-dots served as a vehicle for delivering dyes and drugs into S. cerevisiae cells. The internalized DNA-C-dots and Rh 6G (or Dox) were visualized by green (a and d) and red channels (b and e) of the CFM, respectively. The overlaid images (c and f) revealed the colocalization of DNA-C-dots and Rh 6G, whereas Dox had a distinct punctate distribution, suggesting the release of Dox from the vehicles within the cells. (Insets) Enlarged images. (Reproduced with permission from reference, [65] Copyright 2015 American chemical society)

Surprisingly, DNA-C-dots or Rh 6G displayed an evenly distributed pattern into cells, which is shown by the strong fluorescence signal from the cells whereas the DOX exhibited a punctate pattern suggesting that DNA-C-dots can function as vehicles for drug (DOX) delivery. To support this statement, propidium iodide (PI) was used (instead of Rh 6G and Dox) as a drug molecule that was loaded onto DNA-C-dots by taking advantage of their electrostatic interactions. It is well documented that Rh-6G or Dox can enter the live plasma membrane of eukaryotic cells without the carrier nonselectively via passive diffusion. However, propidium iodide is not permeant to both normal and cancerous cells and thus frequently serves as a marker for dead cell detection. [68] In this study, S. cerevisiae and HEK 293 cells were used for the evaluation of PI delivery function of DNA-C-dots. The obtained data demonstrated that DNA-C-dots can not only act as a carrier in the plasma membrane but also transport PI in and out of the cells under some circumstances. Recently Dong et al. [69] introduced a novel nanocarrier with nitrogen and phosphorus-based dual heteroatoms doped hollow carbon dots (PNHC-dots) for drug delivery and cell imaging. These heteroatoms doped C-dots show many advantages over pure carbon dots. By introducing defect such as N, P, S etc., chemical activities can be improved, the local chemical properties of doped C-dots can tune the structure of electronic energy levels and it also provides the availability of functional groups for target sensing. It also offers unique optical properties including higher photostability, higher fluorescent quantum yield, less photobleaching, low toxicity, higher biocompatibility and more surface-passivated groups with emerging employment in bioanalytical research. Although, only N-doped C-dots show several drawbacks including lower stability, weak resistance to metal ion interference and high reactivity towards acid or base. Therefore, incorporating larger atoms (P or S) as some dual doped materials can not only have preferable influences on the polarizabilities but also alter electronic characteristics by creating more active sites of carbon-based nanomaterials with unanticipated biological applications. Although, many methods have been developed for the synthesis of nitrogen and phosphorus based co-doped C-dots, [70-72] with amines/ammonia as the N-dopant source and phosphoric acid as the P-dopant source respectively. However, most of synthetic pathways require complicated operations, long reaction times, harsh synthesis conditions, complicated heating processes and tedious workup procedures, including rigorous synthesis temperature. In this regard, a simple automatic bottom-up method was employed for the synthesis of heteroatoms co-doped carbon dots (PNHC-dots) by mixing of glucose, ethylene diamine and phosphoric acid. This synthesis avoids any external heat treatment and is a cost-effective synthetic route with large scale production. Structural characterizations by various spectroscopic methods revealed that nitrogen and phosphorus dual-doped hollow carbon dots (PNHC-dots) possess hollow structure with small oxygenious graphite domains as well as abundant phosphate, hydroxyl, pyridinic, pyridine-like and pyrrole-like nitrogen groups. Therefore, these unique features endow PNHC-dots with remarkable photoluminescent properties, making PNHC-dots more accurate as an optical code for *In vitro* and *In vivo* cell studies. The unique co-doped hollow nanoplatform facilitates the effective loading of anticancer drugs such as Doxorubicin (DOX) and subsequent controlled release at the target site. Due to the inherent fluorescent nature of Doxorubicin, (DOX), this model anticancer drug served as a both therapeutic component and an imaging agent. In this study, DOX was adsorbed onto PNHC-dots via electrostatic interaction that served to form a dual emission drug delivery system (Figure 2). PNHC-dots exhibited sufficient water dispersity in wide pH range of solution and displayed negligible cytotoxicity and excellent biocompatibility. Compared with conventional drug carriers, PNHC-dots-based anticancer nanocarrier for drug possesses many potential advantages including high drug-loading, rapid cellular uptake and pH-controlled drug-release, which is of crucial importance for many biomedical applications. In addition, this PNHC-dots-based anticancer drug nanocarrier is highly effective at inhibiting tumor growth and also improved targeted therapy for cancer treatment. The surface passivated nitrogen and phosphorus dual-doped hollow carbon dots (PNHC-dots) were prepared via a facile and scalable one pot acid-base reaction under thermal conditions. [73] In brief, 1,2-ethylenediamine (EDA) and glucose were dissolved with phosphoric acid (85%) (EDA: glucose: H3PO4 ratio 6.0: 0.4: 4.0). The sticky slurry mixture was expanded due to the foaming of the boiling glucose from the spontaneous heat.

After the mixture changed to brown, the reaction mixture was continued for another 10 minutes until the acid-base reaction was completed. Then the reaction mixture was cooled at room temperature. The obtained sticky solid (PNHC-dots) was neutralized with NaOH. The resulting precipitate was repeatedly washed, followed by a dialysis treatment to remove excess salt. The present synthetic method is quick, simple and can be fabricated using a cost-effective solution method. During this exothermic, self-assembled carbonization and vaporization reaction, the released heat was employed to generate self-engineered nanobubbles from the mixtures of DEA, glucose, phosphoric acid and water. These nanobubbles were used as a template to synthesize PNHC-dots morphology with hollow interiors, which was confirmed by transmission electron microscopy (TEM), atomic force microscopy (AFM), dynamic light scattering (DLS) and other spectroscopic techniques.

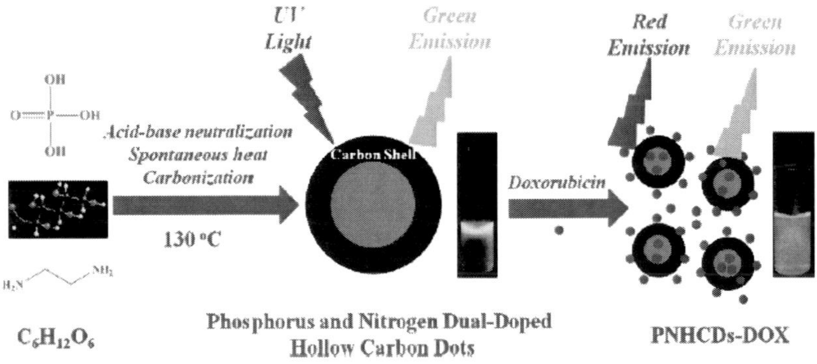

Figure 3. Schematic representation of one-step fabrication of luminescent PNHCDs and the DOX loading process. (Reproduced with permission from reference, [69] Copyright 2016 American Chemical Society)

The aqueous PNHC-dots solution possesses two characteristic absorption bands at 275 nm and 336 nm. The absorption band at 275 nm attributed to the n → π* transition of the C=O band and a prominent absorption peak at 336 nm, which can be ascribed to the formation of excited defect surface states induced by the N and P heteroatoms. [74, 75] In addition, the obtained dispersed PNHC-dots are yellow under day light irradiation but displayed strong green emission upon illumination with UV-light (365 nm) with a high quantum yield. To further explore photoluminescence (PL) properties, the PNHC-dots were tuned by varying the excitation wavelength from 350 to 510 nm. The maximum emission peaks exhibit a red shifted and tunable across the entire visible region from 400 nm (violet emission) to 575 nm (green emission). This corresponds to the non-uniform particle size and distribution of different surface states. This excitation-dependent multicolor typical behavior may be employed in biological research, such as *In vitro*/*In vivo* drug delivery and imaging (vide infra). In addition, fluorescence measurements reveal that there are no significant changes of the PL intensity, peak characteristics as well as sensing stability at different ionic strengths, PBS buffer, cell medium, human serum, and protein standard solution. In this aspect, PNHC-dots can be beneficial because it is necessary to maintain a similar physical environment for practical *In vivo* and *In vitro* testing. The emission intensities decrease significantly at pH values <3, and at pH values >10.0, but almost remain constant over wide ranges of pH values (5−10), indicating its potential application for cell treatment. Moreover, PNHC-dots exhibit superior photo stability and resistance towards photobleaching upon continuous exposure to UV irradiation for several hours. The powder PNHC-dots do not aggregate after subjected to repeated re-dispersing in water which could be beneficial for preservation and transportation. These dots seem to be well dispersed with uniform sizes and exhibit a long term homogenous phase without any precipitation at ambient conditions with a high colloidal stability. Due to their hollow nanostructure and extraordinary PL properties with huge surface bearing capacity, PNHC-dots can host large quantities of drugs with the virtue of acting as drug nanocarriers, which will endow them to internalize and withstand considerable drug loading capacity. Drug loading and delivery studies were conducted by simply mixing of Doxorubicin (DOX) with PNHC-dots solution, where the PNHC-dots acts as a drug nanocarrier system. The drug loading percentage could end up being very high [76] due to noncovalent conjugation through passive adsorption of DOX onto PNHC-dots. The homogeneous aqueous solution of PNHC-dots-DOX complex is very stable and displayed pink luminescence under UV irradiation without any noticeable precipitation for several weeks. The DOX-PNHC-dots complex was confirmed by differences in the absorption spectra, fluorescence measurements and zeta potential surface charge analysis. Due to multitudinous electrostatic interactions, such as π−π stacking through conjugated aromatic rings, CH−π interactions, Vander wall forces and hydrogen bonding between hydroxyl/phosphate groups of PNHC-dots with the hydroxyl/amine groups of DOX, between the dissociated surface phosphate groups of PNHC-dots and the protonated amine groups of DOX, the drug loading capacity is very high compared to un-doped carbon dots. [77,78] In addition, inner empty space in the hollow cavity of PNHC-dots can provide an interactive site with a greater area, which can efficiently accommodate more drug molecules. [79] Generally, drugs preferred to regulate loading and releasing processes in familiar environments. The acidic physiological environment of intracellular lysosomes that enabled high extracellular drug loading led to the release of the loading drug (DOX) inside the cancer cells to cause cell death. Figure 2 represents the fabrication of hetero atom co-doped PNHC-dots loaded with DOX drug molecules as a PNHC-dots-DOX complex via one step process. To investigate the release rate of doxorubicin (DOX) from the drug-delivery system (PNHC-dots-DOX), the solution was transferred in a dialysis membrane tube (MWCO 3500D) and the membrane was immersed in a vial containing 25 ml of phosphate buffered saline (PBS) solution with gentle stirring at 37oC and pH = 7. It is well documented that the pH value of the releasing medium can trigger a higher drug release rate from different drug loaded carbon dots as nanocarriers with active functional groups such as carboxylic/phosphoric acid, ammonia or amines etc. [80] Considering the pH sensitive, surface-active functional groups, drug release kinetics studies are routinely performed to examine the drug (DOX) release ability of PNHC-dots-DOX *In vitro* in PBS buffer solution at different pH values to simulate the endosomal and extracellular environments, respectively. [81] Nevertheless, the DOX was hardly released from PNHC-dots-DOX (only ca. 3%) at pH 7.4 during the initial 3 h. When the pH of the medium was adjusted to pH 5.0, there was a burst of DOX release (nearly 96% within 3 h) from the PNHC-dots-DOX complex, suggesting that the more acidic environment led to the release of the drug at a much faster rate compared to neutral condition. [82] In this regard, the acidic pH in the extracellular tumor microenvironment [83] (pH 6-7) as well as the intracellular endosomes and lysosomes (pH 4-6), can be used as some external stimuli to trigger that potentially facilitating DOX release form the PNHC-dots-DOX complex. To understand the possible mechanism for the faster drug release rate at pH 5.0 than at pH 7.4, it was hypothesized that due to protonation of amine groups in DOX molecules under acidic environment, the water solubility of DOX molecules is higher than the neutral pH. The phosphoric acids anchored on PNHC-dots dissociate as a phosphate backbone with negative charges at neutral pH along with the amphipathic molecule DOX contained amine groups that hold some positive charges. Therefore, DOX molecules can strongly interact with the carrier PNHC-dots through hydrogen bonding and other noncovalent interactions. The negative charge of PNHCDs become weakened at lower of pH (pH 5.0), due to incomplete dissociation of the phosphoric acid moieties. However, at the same pH, the completely dissociated protonated amine of the DOX molecules can increase the hydrophilicity or solubility significantly in aqueous solutions, which indicates that the DOX release from PNHC-dots-DOX complex more easily in acidic medium (pH 5.0) than in neutral medium (pH 7.4). These results show that the targeted PNHC-dots-DOX complex possesses high drug association and sustained drug release properties apposite for targeted tumor treatment that can be widely used in clinical practice with improved drug delivery and cure rate and reduce adverse drug reactions. [84, 85]

To use the PNHC-dots as drug delivery vehicles, the cytotoxicity of DOX, PNHC-dots and PNHC-dots-DOX, were evaluated with three cancer cells (HepG2, SiHa, and MCF-7) and three noncancerous (L-02, Ect 1, and MCF-10A) for comparison. As shown in Figure 3a-f, all the cell lines, HepG2 (Figure 3a), SiHa (Figure 3b), MCF-7 (Figure 3c), L-02 (Figure 3d), Ect 1 (Figure 3e), MCF-10A (Figure 3f) were incubated with various concentrations of PNHC-dots, DOX, and PNHC-dots-DOX ranging from ca. 10 μg mL−1 to 800 μg mL−1 for 48 h. The MTT assay was used to evaluate the cell viabilities of all cell lines. After 48 h of treatment, there was no obvious toxic influence from the PNHC-dots for the cell growth or cell proliferation within the tested PNHC-dots doses ranges, and nearly 100% were found to be healthy after extended incubation up to 96 hours at similar concentration ranges (Figure 3). Though, the free DOX was prone to be highly inimical and displayed significant cytotoxicity on the mentioned cell lines at investigated concentration ranges. Cell viabilities of all kinds of tested cells were significantly decreased to nearly 13−20% with increasing DOX doses from 50 to 800 μg mL−1. It is worth mentioning that free DOX is more effective to destroy cancer cells with adverse side effects in cancer patients at optimal concentration, however PNHC-dots-DOX could able to inhibit cell proliferation as well as reduce tumor growth significantly with negligible side effects. [86] At same concentrations of free DOX and DOX loaded PNHC-dots with all types of tested cell treatments caused dose-dependent death with similar trends. For example, when the concentrations of DOX and DOX loaded PNHC-dots are about 400 μg mL−1, dose-dependent death was observed 73−77% and 65−70%, respectively, upon treatment with all tested cell lines. Although PNHCDs do not cause any interference in the cell survival, the loaded DOX drugs show negligible toxicity, [87] which is only due to the individual influences of free DOX. To perform the control experiment, similar dosages of DOX loaded PNHC-dots and free DOX were used for the treatment of all three healthy cells. The results indicate that combination treatment (DOX loaded PNHC-dots) induced much less cytotoxicity than the normal healthy cells at a similar dosage. This controlled drug release system (DOX loaded PNHC-dots) can efficiently enhance the intracellular drug delivery, allow for sustained drug release for extend period, infer their potential use for targeting of tumor cells, image the tumor microenvironment and improve cancer therapy *In vivo* and *In vitro*. To examine the localization of DOX loaded PNHC-dots into cells through rapid internalization *In vitro*, a laser scanning confocal microscopy (LSCM) was used to image specific features of SiHa cells incubated with PNHC-dots and DOX loaded PNHC-dots, which is a crucial aspect of a drug delivery system. Figure 3g represents the fluorescent images of SiHa cells with DOX loaded PNHC-dots under different excitation wavelengths (3gi and 3gii) and their overlap image (3giii). Figure 3h signifies the bright-field (3hi) and fluorescent images of SiHa cells with free DOX under 543 nm excitation filter (3hii) and their overlaid image (3hiii). The SiHa cells labeled with DOX loaded PNHC-dots exhibit green fluorescence under blue (488) light illumination. After 5 hours incubation under dark, a noticeable green emission was displayed (Figure 3gi) only from the cell membrane and cytoplasm, whereas no infiltration was observed into the cell nuclei. This observation confirms that PNHC-dots form DOX loaded PNHC-dots can be internalized into cells via endocytosis and easily penetrated the cell cytoplasm but did not enter the cell nuclei. However, when the incubation time was longer than 5 hours, DOX had started to penetrate the nucleus, released DOX molecules and displayed red emission from nucleus region (Figure 3gi) under green laser at 543 nm. In addition, single cell incubation results indicate that fluorescence localization (basically intensity emitted from the cell) was distributed uniformly over the whole cell's surface. [88] When the incubation time was extended over 5 hours, there was some negligible changes in the cell intensities from the cytoplasm region where PNHC-dots were dispersed and went in and out of a cell and vice versa. In contrast, the fluorescence intensity of DOX molecule dramatically reduced from cell nuclei [89] or quenched due to intercalation of DOX into the DNA or by self-quenching that can occur at high concentrations of molecularly dissolved DOX. [90] The resultant reduced fluorescence signal data indicated that DOX was released from PNHC-dots-DOX complex and would defuse slowly to the cell nuclei over time. The resultant decrease in the emission intensity followed by diffusion confirmed that the pH-triggered release of DOX from the entrapped acidic vesicles to the cytosol and its diffusion to the target site, the nucleus. [91] Similar experiments were conducted with free PNHC-dots in presence of SiHa cells at different incubation period. It is worth noting that free PNHC-dots can be well distributed and well-dispersed to the cytoplasm region as well as cell nuclei when the incubation time was prolonged up to 24 h. However, under the appropriate incubation time (less than 6 hours), PNHC-dots can be internalized by SiHa cells and pass through cell membranes by passive diffusion to enter the cytoplasm but not in the cell nucleus. The findings of this study suggested that the dual doped PNHC-dots exhibit excellent photostability and can be used for a long-term cellular imaging agent. This study also describes the detailed cellular uptake and trafficking mechanisms of DOX loaded PNHC-dots, where PNHC-dots-DOX complexes form vesicles that enter cells via endocytosis through endoplasmic reticulum penetration into the cytosol. Then, the PNHC-dots-DOX-vesicles or and vacuoles are transported into the lysosomes. Finally, under the acidic circumstance of lysosomes and endosomes, DOX is released and then enters and accumulates in the nuclei which is shown in scheme 2. Thus, considering the efficient targeting capability of PNHC-dots-DOX compared to free DOX, higher therapeutic efficacy relative to other nanomaterials, wavelength dependent PL behavior, good stability and anti-interference performance of PNHC-dots-DOX in a complex biological environment, the multifunctional co-doped PNHC-dots may be a potential candidate for bioimaging, drug delivery, tracking, and other biomedical research. [92]

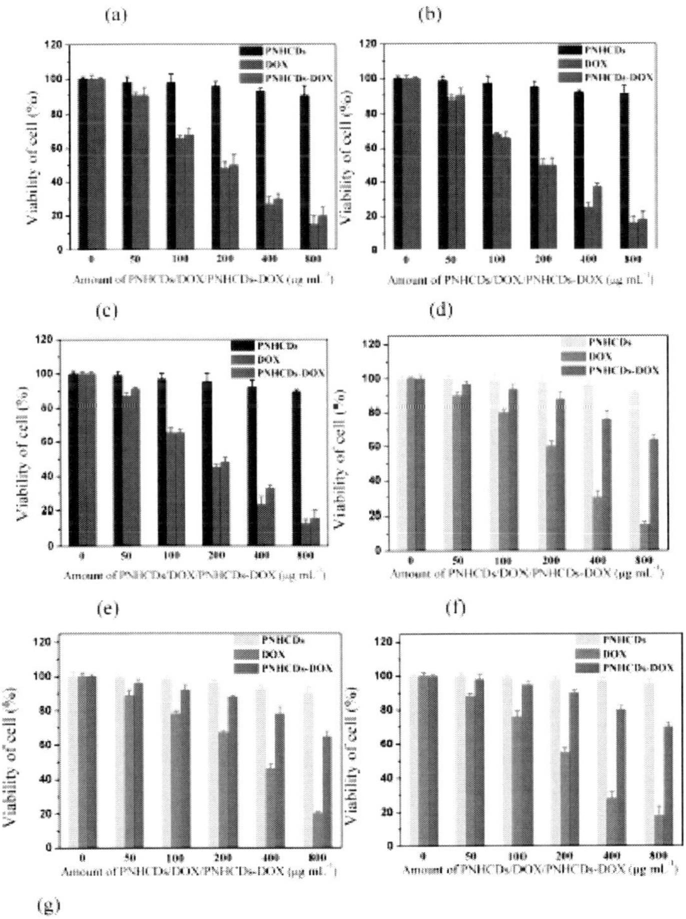

Figure 4. (Continued)

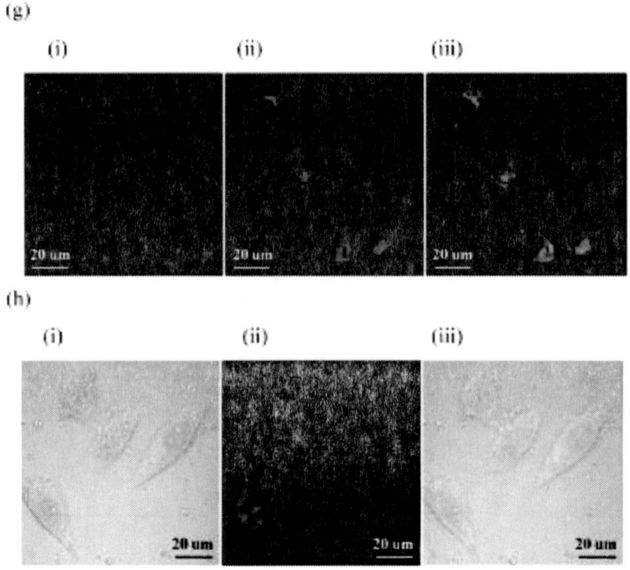

Figure 4. cytotoxicity of PNHCDs, DOX, and PNHCDs-DOX on (a) HepG2, (b) SiHa, (c) MCF-7, (d) L-02, (e) Ect 1, and (f) MCF-10A cell viability. (g) LSCM images of SiHa cells incubated with PNHCDs-DOX (DOX concentration: 0.10 mg mL−1) for 5 h observed under (i) 488/500 ± 25 nm, (ii) 543/650 ± 25 nm, and (iii) their merged images, respectively. (h) LSCM images of SiHa cells incubated with DOX (0.10 mg mL−1) for 3 h observed under (i) bright field, (ii) 543/650 ± 25 nm, and (iii) their merged images, (Reproduced with permission from reference, [69] Copyright 2016 American Chemical Society).

Conclusion and Future Outlook

We summarized the recent advances in the development of different types of targeting moieties with several carbon-based nanomaterial assemblies for targeted drug anticancer drug delivery. All of the described methods for the preparations of multitudinous heteroatom doped carbon dots assemblies, such as the hollow carbon dots, synthesized from Bovine Serum Albumin (BSA) using the solvothermal method, the DNA-C-dots, prepared by a simple carbonization approach via a hydrothermal treatment of Escherichia coli genomic DNA and the nitrogen and phosphorus dual-doped hollow carbon dots (PNHC-dots) prepared via acid–base reaction heat method by the reaction of glucose, ethylene diamine and phosphoric acid, are cost effective, green, simple and reproducible one step convenient processes that represent the potential major advance for industrial scale production. These green synthetic methodologies could be useful tools for developing into effective nanomaterials-based drug delivery diagnostic vehicles today while avoiding intimidating complications. Despite all this, the mentioned heteroatom doped carbon dots from the in-vitro analysis confirmed that they are indeed not only advantageous in terms of low-toxicity, favorable biocompatibility, excellent water solubility, super resistance to photobleaching, surface modification flexibility with sufficient fluorescence quantum yield, they are also highly sensitive as a fluorescence imaging agent for selective target analytes which are comparable to the well-developed, commercially available CdSe/ZnS QDs. We also examined the comparative pH- triggered DOX release efficiencies with three different types of environmentally friendly heteroatom doped carbon dots as novel drug delivery systems *in vitro* at different pH levels that mimic the environment of human body. Due to their hollow core-shell structures, the C-dots have more space to drug load and can accommodate the maximum number of drug with super-high drug loading capacities. We have also elaborately emphasized the cellular and nuclear uptake mechanisms of different types of DOX loaded carbon dots, where C-dot nanocarriers exhibit multiple functionalities, including potential targeted carriers and accelerate the cell nucleus accumulation and enhance the DNA cleavage activities that can be used for targeted cancer therapies. In addition, *In vitro* drug release studies suggested that drug loading with all these heteroatom doped carbon dot complexes were relatively stable at biological pH but gradually releases drug under acidic environment by the dissociation of the complexes. Although, the cellular toxicity and cell viability results suggested that all the heteroatoms doped carbon dots are nontoxic and can be utilized for more suitable drug delivery vehicles, but their clinical applications are of still at the infancy stage.

Overall, all of these heteroatom doped drug nanocarriers not only offers some environmentally friendly synthetic methods, but they also show excellent promise as prospective anticancer agents. The doped carbon dots with nanoscale sizes possess enhancer activities such as to enter the blood circulation system, boosted effective drug encapsulation efficiency, pH responsive targeted drug desorption, no physiochemical changes during endocytosis/loading with nanocarriers, least toxicity to normal cells and other associated organs than other traditional nanomaterials. Furthermore, to practically translate all these heteroatom-doped carbon-based materials as drug delivery vehicles in a clinical setting, several additional issues need to be considered, including total body clearance, scalable low-cost production, environmental impact and more to support therapeutic compliance. As such, drug delivery theranostic nanoplatforms are being constantly developed by combining doped carbon-based materials with several drug molecules. However, despite the research emphasized here, much more work needs to be done to develop safe nanomaterials that can be used as a potential active ingredient in pharmaceutical products.

Acknowledgements

We thank NSF-PREM grant # DMR-1205194, NSF CREST grant # 1547754 and NSF RISE grant # 1547836 for their generous funding.

References

[1] Sun, X., Lei, Y., (2017). Fluorescent carbon dots and their sensing applications. *Trends Anal. Chem.* 89, 163-180.

[2] Igor, L., Uyeda, H. T., Goldman, E. R., (2005). Mattoussi, H. Quantum dot bioconjugates for imaging, labelling and sensing. *Nat. Mater.* 4, 435-446.

[3] Shen, P., Xia, Y., (2014). Synthesis-Modification Integration : One-Step Fabrication of Boronic Acid Functionalized Carbon Dots for Fluorescent Blood Sugar Sensing. *Anal. Chem.* 86, 5323-5329.

[4] Nie, H., Li, M., Li, Q., Liang, S., Tan, Y., Sheng, L., Shi, W., Zhang, S. X-A., (2014). Carbon Dots with Continuously Tunable Full-Color Emission and Their Application in Ratiometric pH. *Sensing Chem. Mater.* 26, 3104-3112.

[5] Sun, H., Wu, L., Wei, W., Qu, X., (2013). Recent advances in graphene quantum dots for sensing. *Mater. Today.* 16, 433-442.

[6] Cao, L., Sahu, S., Anilkumar, P., Bunker, C. E., Xu, J., Fernando, K.A.S., Wang, P., Guliants, E. A., Tackett, K. N., Sun, Y-P., (2011). Carbon nanoparticles as visible-light photocatalysts for efficient CO_2 conversion and beyond. *J. Am. Chem. Soc.* 133, 4754- 4757.

[7] Zhang, Z., Zheng, T., Li, X., Xu, J., Zeng, H., (2016). Progress of Carbon Quantum Dots in Photocatalysis Applications. *Part. Part. Syst. Charact.* 33, 457-472.

[8] 8.Yu, H., Shi, R., Zhao, Y., Waterhouse, G. I. N., Wu, L-Z., Tung, C-H., Zhang, T., (2016). Smart Utilization of Carbon Dots in Semiconductor Photocatalysis. *Adv. Mater.* 28, 9454-9477.

[9] Wang, R., Lu, K-Q., Tang, Z-R., Xua, Y-J., (2017). Recent progress in carbon quantum dots : synthesis, properties and applications in photocatalysis. *J. Mater. Chem. A.* 5, 3717-3734.

[10] Xie, Z., Yin, Z., Wu, Y., Liu, C., Hao, X., Du, Q., Xu, X., (2017). White Light-Emitting Diodes Based on Individual Polymerized Carbon Nanodots. *Sci. Rep.* 7, 12146.

[11] Sekiya, R., Uemura, Y., Murakami, H., Haino, T., (2014). White-Light-Emitting Edge-Functionalized Graphene Quantum Dots. *Angew. Chem.* 126, 5725-5729.

[12] Zhang, Xi., Zhang, Y., Wang, Y., Kalytchuk, S., Kershaw, S. V., Wang, Y., Wang, P., Zhang, T., Zhao, Y., Zhang, H., Cui, T., Wang, Y., Zhao, J., Yu, W. W., Rogach, A. L., (2013). Color-Switchable Electroluminescence of Carbon Dot Light-Emitting Diodes. *ACS Nano* 7, 11234-11241.

[13] Zhang, H., Yee, D., Wang, C., (2008). Quantum dots for cancer diagnosis and therapy : biological and clinical perspectives. *Nanomedicine* 3, 83-91.

[14] Mahmoudi, M., Hosseinkhani, H., (2011). Hosseinkhani, M., Boutry, S., Simchi, A., Journeay, W. S., Subramani, K., Laurent, S., Magnetic resonance imaging tracking of stem cells *in vivo* using iron oxide nanoparticles as a tool for the advancement of clinical regenerative medicine. *Chem. Rev.* 111, 253-280.

[15] Zhang, J., Yu, S-H., (2016). Carbon dots : large-scale synthesis, sensing and bioimaging. *Mater. Today* 19, 382-393.

[16] Sanvicens, N., Marco, M. P., (2008). Multifunctional nanoparticles–properties and prospects for their use in human medicine. *Trends Biotechnol.* 26, 425-433.

[17] Luo, P. G., Sahu, S., Yang, S-T., Sonkar, S. K., Wang, J., Wang, H., LeCroy, G. E., Cao, L., Sun, Y-P., (2013). Carbon "quantum" dots for optical bioimaging *J. Mater. Chem. B* 1, 2116-2127.

[18] Ho, D., Wang, C-H K., Chow, E. K-H., (2015). Nanodiamonds : The intersection of nanotechnology, drug development, and personalized medicine. *Sci Adv.* 1, e1500439.

[19] Saito, N., Haniu, H., Usui, Y., Aoki, K., Hara, K., Takanashi, S., Shimizu, M., Narita, N., Okamoto, M., Kobayashi, S., Nomura, H., Kato, H., Nishimura, N., Taruta, S., Endo, M., (2014). Safe Clinical Use of Carbon Nanotubes as Innovative Biomaterials. *Chem. Rev.* 114, 6040-6079.

[20] Zrazhevskiy, P., Gao, X., (2009). Multifunctional quantum dots for personalized medicine, *Nano Today* 4, 414-428.

[21] Rizvi, S. B., Ghaderi, S., Keshtgar, M., Seifalian, A. M., (2010). Semiconductor quantum dots as fluorescent probes for *in vitro* and *in vivo* bio-molecular and cellular imaging. *Nano Reviews* 1, 5161.

[22] Li, X., Rui, M., Song, J., Shen, Z., Zen, H., (2015). Carbon and Graphene Quantum Dots for Optoelectronic and energy devices: A review. *Adv. Funct. Mater.* 25, 4929-4947.

[23] Peng, J., Gao, W., Gupta, B. K., Liu, Z., Romero-Aburto, R., Ge, L., Song, L., Alemany, L. B., Zhan, X., Gao, G., Vithayathil, S. A., Kaipparettu, B. A., Marti, A. A., Hayashi, T., Zhu, J-J., Ajayan, P. M., (2012). Graphene Quantum Dots Derived from Carbon Fibers. *Nano Lett.* 12, 844-849.

[24] Armaković, S., Armaković, S. J., Koziel, S., (2017). Optoelectronic Properties of Curved Carbon Systems. *Carbon* 111, 371-379.

[25] Sahu, S., Behera, B., Maiti, T. K., Mohapatra, S., (2012). Simple one-step synthesis of highly luminescent carbon dots from orange juice : application as excellent bio-imaging agents. *Chem. Commun.* 48, 8835-8837.

[26] Shen, C., Wang, J., Cao, Y., Lu, Y., (2015). Facile access to B-doped solid-state fluorescent carbon dots toward light emitting devices and cell imaging agents, *J. Mater. Chem. C* 3, 6668-6675.

[27] Song, Y., Zhu, C. Song, J., Li, H., Du, D., Lin, Y., (2017). Drug-Derived Bright and Color-Tunable N-Doped Carbon Dots for Cell Imaging and Sensitive Detection of Fe^{3+} in Living Cells. *ACS Appl. Mater. Interfaces* 9, 7399-7405.

[28] Zhao, S., Lan, M., Zhu, X., Xue, H., Ng, T-W., Meng, X., Lee, C-S., Wang, P., Zhang, W., (2015). Green Synthesis of Bifunctional Fluorescent Carbon Dots from Garlic for Cellular Imaging and Free Radical Scavenging. *ACS Appl. Mater. Interfaces* 7, 17054-17060.

[29] Wang, Y., Anilkumar, P., Cao, L., Liu, J-H., Luo, P. G., Tackett II, K. N., Sahu, S., Wang, P., Wang, X., Sun, Y-P., (2011). Carbon dots of different composition and surface functionalization : cytotoxicity issues relevant to fluorescence cell imaging. *Experimental Biology and Medicine* 236, 1231-1238.

[30] Pramanik, A. Jones, S., Pedraza, F., Vangara, A., Sweet, C., Williams, M. S., Ruppa-Kasani, V., Risher, S. E., Sardar, D., Ray, P. C., (2017). Fluorescent, Magnetic Multifunctional Carbon Dots for Selective Separation, Identification, and Eradication of Drug-Resistant Superbugs. *ACS Omega* 2, 554-562.

[31] Pramanik, A., Vangara, A., Nellore, B. P. V., Sinha, S. S., Chavva, S. R., Jones, S., Ray, P. C., (2016). Development of Multifunctional Fluorescent–Magnetic Nanoprobes for Selective Capturing and Multicolor Imaging of Heterogeneous Circulating Tumor Cells. *ACS Appl. Mater. Interfaces* 8,15076-15085.

[32] Wang, Q., Huang, X., Long, Y., Wang, X., Zhang, H., Zhu, R., Liang, L., Teng, P., Zheng, H., 2013. Hollow luminescent carbon dots for drug delivery, *Carbon*, 59, 192-199.

[33] Ding, H., Zhang, F., Zhao, C., Lv, Y., Ma, G., Wei, W., Tian, Z., (2017). Beyond a Carrier : Graphene Quantum Dots as a Probe for Programmatically Monitoring Anti-Cancer Drug Delivery, Release, and Response. *ACS Appl. Mater. Interfaces*. 9, 27396-27401.

[34] Feng, T., Ai, X., An, G., Yang, P., Zhao, Y., (2016). Charge-Convertible Carbon Dots for Imaging-Guided Drug Delivery with Enhanced *In vivo* Cancer Therapeutic Efficiency. *ACS Nano 10*, 4410-4420.

[35] Ding, H., Du, F., Liu, P., Chen, Z., Shen, J., (2015). DNA–Carbon Dots Function as Fluorescent Vehicles for Drug Delivery. *ACS Appl. Mater. Interfaces* 7, 6889-6897.

[36] Yao, Y-Y., Gedda, G., Girma, W. M., Yen, C-L., Ling, Y-C., Chang, J-Y., (2017) Magnetofluorescent Carbon Dots Derived from Crab Shell for Targeted Dual-Modality Bioimaging and Drug Delivery. *ACS Appl. Mater. Interfaces*, 9, 13887-13899.

[37] Pardo, J., Peng, Z., Leblanc, R. M., (2018) Cancer Targeting and Drug Delivery Using Carbon-Based Quantum Dots and Nanotubes. *Molecules* 23, 378.

[38] Zhao, M-X., Zhu, B-J., (2016). The Research and Applications of Quantum Dots as Nano-Carriers for Targeted Drug Delivery and Cancer Therapy. *Nanoscale Res. Lett.* 11, 207.

[39] D., Chen. Gao, H., Chen, X., Fang, G., Yuan, S., Yuan, Y., (2017). Excitation-Independent Dual-Color Carbon Dots : Surface-State Controlling and Solid-State Lighting. *ACS Photonics* 4, 2352-2358.

[40] Ding, H., Yu, S-B., Wei, J-S., Xiong, H-M., (2016). Full-Color Light-Emitting Carbon Dots with a Surface-State-Controlled Luminescence Mechanism. *ACS Nano* 10, 484-491.

[41] Liu, J-H., Cao, L., LeCroy, G. E., Wang, P., Meziani, M. J., Dong, Y., Liu, Y., Luo, P. G., Sun, Y-P., (2015). Quantum Dots for Fluorescence Labeling of Cells, *ACS Appl. Mater. Interfaces Carbon* 7, 19439-19445.

[42] Liu, Q., Guo, B., Rao, Z., Zhang, B., Gong, J. R., (2013). Strong Two-Photon-Induced Fluorescence from Photostable, Biocompatible Nitrogen-Doped Graphene Quantum Dots for Cellular and Deep-Tissue Imaging. *Nano Lett.* 13, 2436-244.

[43] Xu, X., Ray, R., Gu, Y., Ploehn, H. J., Gearheart, L., Raker, K., Scrivens, W. A., (2004). Electrophoretic Analysis andPurification of Fluorescent Single-Walled Carbon Nanotube Fragments. *J. Am. Chem. Soc.* 126, 12736-12737.

[44] Sun, Y-P., Zhou, B., Lin, Y., Wang, W., Fernando, K. S., Pathak, P., Meziani, M. J., Harruff, B. A., Wang, X., Wang, H., (2006). Quantum-Sized Carbon Dots for Bright and Colorful Photo-luminescence. *J. Am. Chem. Soc.* 128, 7756-7757.

[45] Bazaka, K., Jacob, M. V., Ostrikov, K., (2016). Sustainable Life Cycles of Natural-Precursor-Derived Nanocarbons. *Chem. Rev.* 116, 163-214.

[46] Parent, P., Laffon, C., Marhaba, I., Ferry, D., Regier, T. Z., Ortega, I. K., Chazallon, B., Carpentier, Y., Focsa, C., (2016) Nanoscale characterization of aircraft soot : A high-resolution transmission electron microscopy, Raman spectroscopy, X-ray photoelectron and near-edge X-ray absorption spectroscopy study. *Carbon* 101, 86-100.

[47] Hu, S., (2016). Tuning Optical Properties and Photocatalytic Activities of Carbon based “Quantum Dots” Through their Surface Groups. *Chem. Rec.* 16, 219-230.

[48] Liu, H., He, Z., Jiang, L-P., Zhu, J-J., (2015). Microwave-Assisted Synthesis of Wavelength-Tunable Photoluminescent Carbon Nanodots and Their Potential Applications. *ACS Appl. Mater. Interfaces* 7, 4913-4920.

[49] Xu, Y., Liu, J., Gao, C., (2014). Erkang Wang Applications of carbon quantum dots in electrochemiluminescence : A mini review. *Electrochem commun.* 48, 151-154.

[50] Jiang, K., Sun, S., Zhang, L., Wang, Y., Cai, C., Lin, H., (2015). Bright-Yellow-Emissive N-Doped Carbon Dots: Preparation, Cellular Imaging, and Bifunctional Sensing. *ACS Appl. Mater. Interfaces* 7, 23231-23238.

[51] Li, Y., Lin, H., Luo, C., Wang, Y., Jiang, C., Qi, R., Huang, R., Travas-sejdic, J., Peng, H., (2017). Aggregation induced red shift emission of phosphorus doped carbon dots, *RSC Adv.* 7, 32225-32228.

[52] Barman, M. K., Jana, B., Bhattacharyya, S., Patra, A., (2014). Photophysical Properties of Doped Carbon Dots (N, P, and B) and Their Influence on Electron/Hole Transfer in Carbon Dots–Nickel (II) Phthalocyanine Conjugates, *J. Phys. Chem. C* 118, 20034-20041.

[53] Xu, Q., Pu, P., Zhao, J., Dong, C., Gao, C., Chen, Y., Chen, J., Liu, Y., Zhou, H., (2015). Preparation of highly photoluminescent sulfurdoped carbon dots for Fe(III) detection, *J. Mater. Chem. A* 3, 542-546.

[54] Wang. J., Liu, G., Leung, K. C., Loffroy. R., Lu, P. X., Wang, Y. X., (2015). Opportunities and Challenges of Fluorescent Carbon Dots in Translational Optical Imaging. *Curr Pharm Des.* 21, 5401-5416.

[55] Luo, P. G., Yang, F., Yang, S-T., Sonkar, S. K., Yang, L., Broglie, J. J., Liu, Y., Sun, Y-P. (2014). Carbon-based quantum dots for fluorescence imaging of cells and tissues. *RSC Adv.* 4, 10791-10807.

[56] Gustafson, H. H., Holt-Casper, D., Grainger, D. W., Ghandehari, H., (2015). Nanoparticle Uptake : The Phagocyte Problem. *Nano Today* 10, 487-510.

[57] Kettiger, H., Schipanski, A., Wick, P., Huwyler, J., (2013). Engineered nanomaterial uptake and tissue distribution : from cell to organism, *Int J Nanomedicine* 8, 3255-3269.

[58] Chithrani, B. D., Ghazani, A. A., Chan, W. C., (2006). Determining the size and shape dependence of gold nanoparticle uptake into mammalian cells. *Nano Lett.* 6, 662-668.

[59] Chithrani, B. D., Chan, W. C., (2007). Elucidating the mechanism of cellular uptake and removalof protein-coated gold nanoparticles of different sizes and shapes. *Nano Lett.* 7, 1542-1550.

[60] Koping-Hoggard, M., Melnikova, Y, S., Varum, K. M., Lindman, B., Artursson, P., (2003). Relationship between the physical shape and the efficiency of oligomeric chitosan as a gene delivery system *In vitro* and *In vivo*. *J Gene Med.* 5,130-141.

[61] Wang, Q., Huang, X., Long, Y., Wang, X., Zhang, H., Zhu, R., Liang, L., Teng, P., Zheng, H., (2013). Hollow luminescent carbon dots for drug delivery. *Carbon* 59, 192-199.

[62] Regev, R., Eytan, G. D., (1997). Flip-flop of doxorubicin across erythrocyte and lipid membranes. *Biochem Pharmacol.* 15, 1151-1158.

[63] Yuan, Y., Guo, B., Hao, L., Liu, N., Lin, Y., Guo, W., Li, X., Gu, B., (2017). Doxorubicin-loaded environmentally friendly carbon dots as a novel drug delivery system for nucleus targeted cancer therapy. *Colloids and Surfaces B : Biointerfaces* 159, 349-359.

[64] Gong, T., Xie, J., Liao, J., (2015). Nanomaterials and bone regeneration. *Bone. Res.* 3, 1-7.

[65] Ding, H., Du, F., Liu, P., Chen, Z., Shen, J., (2015). DNA−Carbon Dots Function as Fluorescent Vehicles for Drug Delivery. *ACS Appl. Mater. Interfaces* 7, 6889-6897.

[66] Zhang, L., Foxman, B., Gilsdorf, J. R., Marrs, C. F., (2005). Bacterial Genomic DNA Isolation Using Sonication for Microarray Analysis. *Bio Techniques* 39, 640-644.

[67] Swietach. P., Vaughanjones, R. D., Harris, A. L., (2014). The chemistry, physiology and pathology of pH in cancer. *Phil. Trans. R. Soc. B* 369, 20130099.

[68] Cummings, B. S., Schnellmann, R. G., (2004). Measurement of Cell Death in Mammalian Cells. *Curr Protoc Pharmacol.* 25, 12.8.1-12.8.22.

[69] Gong, X., Zhang, Q., Gao, Y., Shuang, S., Choi, M. M. F., Dong, C., (2016). Phosphorus and Nitrogen Dual-Doped Hollow Carbon Dot as a Nanocarrier for Doxorubicin Delivery and Biological Imaging. *ACS Appl. Mater. Interfaces* 8,11288-11297.

[70] Sun, X., Brückner, C., Lei, Y., (2015). One-pot and ultrafast synthesis of nitrogen and phosphorus co-doped carbon dots possessing bright dual wavelength fluorescence emission. *Nanoscale* 7, 17278-17282.

[71] Ananthanarayanan, A., Wang, Y., Routh, P., Sk, M. A., Than, A., Lin, M., Zhang, J., Chen, J., Sun, H., Chen, P., (2015). Nitrogen and

phosphorus co-doped graphene quantum dots : synthesis from adenosine triphosphate, optical properties, and cellular imaging. *Nanoscale* 7, 8159-8165.

[72] Shi, B., Su, Y., Zhang, L., Huang, M., Liu, R., Zhao, S., (2016). Nitrogen and Phosphorus Co-Doped Carbon Nanodots as a Novel Fluorescent Probe for Highly Sensitive Detection of Fe^{3+} in Human Serum and Living Cells. *ACS Appl. Mater. Interfaces* 8, 10717-10725.

[73] Bhunia, S. K., Saha, A., Maity, A. R., Ray, S. C., (2013). Jana, N. R. Carbon Nanoparticle-based Fluorescent Bioimaging Probes. *Sci. Rep.* 3, 1473.

[74] Anilkumar, P., Wang, X., Cao, L., Sahu, S., Liu, J-H., Wang, P., Korch, K., Tackett II, K. N., Parenzan, A., Sun, Y-P., (2011). Toward quantitatively fluorescent carbon-based "quantum" dots. *Nanoscale* 3, 2023-2027.

[75] Wang, L., Cao, S.-T., Yang, F., Lu, M. J., Meziani, L., Tian, K. W., Sun, W., Bloodgood, M. A., Sun, Y.-P., (2010). Bandgap-Like Strong Fluorescence in Functionalized Carbon Nanoparticles. *Angew. Chem. Int. Ed.* 49, 5310-5314.

[76] Sun, T., Zheng, M., Xie, Z., Jing, X., (2017). Supramolecular hybrids of carbon dots with doxorubicin : Synthesis, stability and cellular trafficking. *Mater. Chem. Front.* 1, 354-360.

[77] Zhang, M., Yuan, P., Zhou, N., Su, Y., Shao, M., Chi, C., (2017). pH-Sensitive N-doped carbon dots–heparin and doxorubicin drug delivery system : preparation and anticancer research. *RSC Adv.* 7, 9347-9356.

[78] Kumar, J. P., Konwarh, R., Kumar, M., Gangrade, A., Mandal, B. B., (2018). Potential Nanomedicine Applications of Multifunctional Carbon Nanoparticles Developed Using Green Technology. *ACS Sustainable Chem. Eng.* 6, 1235-1245.

[79] Jiao, Y., Sun, Y., Chang, B., Lu, D., Yang, W., (2013). Redox- and Temperature-Controlled Drug Release from Hollow Mesoporous Silica Nanoparticles. *Chem. Eur. J.* 19, 15410-15420.

[80] Meng, F., Zhong, Y., Cheng, R., Deng, C., Zhong, Z., (2014). pH Sensitive Polymeric Nanoparticles for Tumor-Targeting Doxorubicin Delivery : Concept and Recent Advances. *Nanomedicine* 9, 487-499.

[81] Dadsetan, M., Taylor, K. E., Yong, C., Bajzer, Ž., Lu, L., Yaszemskia, M. J., (2013). Controlled release of doxorubicin from pH-responsive microgels. *Acta Biomater.* 9, 5438-5446.

[82] Yoshida, T., Lai, T. C., Kwon, G. S., Sako, K., (2013). pH- and ion-sensitive polymers for drug delivery. *Expert Opin Drug Deliv.* 10, 1497-1513.

[83] Kato, Y., Ozawa, S., Miyamoto, C., Maehata, Y., Suzuki, A., Maeda, T., Baba, Y., (2013). Acidic extracellular microenvironment and cancer. *Cancer Cell Int.* 13, 89.

[84] Tiwari, G., Tiwari, R., Sriwastawa, B., Bhati, L., Pandey, S., Pandey, P., Bannerjee, S. K., (2012). Drug delivery systems : An updated review. *Int J Pharm Investig.* 1, 2-11.

[85] Singh, R., Lillard, J. W., (2009). Nanoparticle-based targeted drug delivery. *Exp Mol Pathol.* 86, 215-223.

[86] Wang, Y., Xie, Y., Li, J., Peng, Z-H., Sheinin, Y., Zhou, J., Oupický, D., (2017). Tumor-Penetrating Nanoparticles for Enhanced Anticancer Activity of Combined Photodynamic and Hypoxia-Activated Therapy. *ACS Nano* 11, 2227-2238.

[87] Ono, R. J., Lee, A. L. Z., Voo, Z. X., Venkataraman, S., Koh, B. W., Yang, Y. Y., Hedrick, J. L., (2017). Biodegradable Strain-Promoted Click Hydrogels for Encapsulation of Drug-Loaded Nanoparticles and Sustained Release of Therapeutics. *Biomacromolecules* 18, 2277-2285.

[88] Zhang, Y., Han, L., Zhang, Y., Chang, Y-Q., Chen, X-W., He, R-H., Shu, Y., Wang, J-H., (2016). Glutathione-mediated mesoporous carbon as a drug delivery nanocarrier with carbon dots as a cap and fluorescent tracer. *Nanotechnology* 27, 5102.

[89] Wang, C., Wu, C., Zhou, X., Han, T., Xin, X., Wu, J., Zhang, J., Guo, S., (2013). Enhancing Cell Nucleus Accumulation and DNA Cleavage Activity of Anti-Cancer Drug via Graphene Quantum Dots. *Sci Rep.* 3, 2852.

[90] Mohan, P., Rapoport, N., (2010). Doxorubicin as a molecular nanotheranostic agent : effect of doxorubicin encapsulation in micelles or nanoemulsions on the ultrasound-mediated intracellular delivery and nuclear trafficking. *Mol. Pharmaceutics* 7, 1959-1973.

[91] Bronk, S. F., Gores, G. J., (1991). Efflux of protons from acidic vesicles contributes to cytosolic acidification of hepatocytes during ATP depletion. *Hepatology* 14, 626-33.

[92] Zrazhevskiy, P., Sena, M., Gaoa, X., (2010). Designing multi-functional quantum dots for bioimaging, detection, and drug delivery. *Chem Soc Rev.* 39, 4326-4354.

In: Nanotechnology
Editor: Paresh Chandra Ray
ISBN: 978-1-53613-889-4

Chapter 3

ENGINEERED PLASMONIC NANOSTRUCTURES FOR BIOLOGICAL SENSING

Anant K. Singh*, Aysha Evans, Sandra L. Barnes and Yolanda K. Jones
Department of Chemistry and Physics, Alcorn State University, Alcorn State, Mississippi, US

ABSTRACT

Cancer is a real problem of our century and one of the leading causes of death, accounting for one of eight deaths occurring worldwide. Based on the actual data of International Agency for Research on Cancer (IARC) estimates ~13.1 million deaths associated to cancer by 2030 [1, 2]. It is becoming clear for many researchers that the low survival rate is due to the lack of adequate drug diagnostics and drug delivery systems and not due to the lack of potent, natural, or synthetic anti-tumoral agents. Therefore, there is a real need to develop more sensitive techniques, drug carriers and delivery systems which would be able to sense and deliver

* Corresponding Author Email: asingh@alcorn.edu.

the therapeutics, only at the specific target site and improve the efficiency of treatment and consequently limiting the unwanted systemic side effects [3, 4]. Plasmonic nano-structures are particularly appealing in the area of medical diagnostics because of their plasmonic optical properties that can be exploited for both diagnostic and therapeutic purpose along with their established bio-compatibility and ease of surface modification. In this chapter, we review a brief recent progress in the various ways to probe the special properties of plasmonic nanostructure for biological diagnostics and therapeutics. Frame of the chapter begins with a brief survey on the available ways to prepare plasmonic nanostructures and associated advantages and disadvantages. Since the influence of size, shape, surface properties, materials characteristics are extremely important factors for a smart sensor design from the biological points of view. We discuss some of the pioneering methods to alter the surface properties by loading functional groups for specific biological sensing and high through put performance. We have also discussed the various highly sensitive spectroscopic techniques specially, colorimetric, fluorescence, Raman and Hyper Rayleigh scattering spectroscopic properties of plasmonic nanostructures to use as a probe to identify the targeted biological pathogens. Achieved level of sensitivity to sense bio-pathogens using these optical techniques has been categorically discussed. Special attention is paid for the benefits and the risk of using any of these spectroscopic methods to use plasmonic structures as a nanoprobes for the detection and monitoring of targeted pathogenic molecule.

Keywords: nanostructures, conjugation, luminescence, Raman

INTRODUCTION

Since the discovery of nanostructure one of the attractive notice was the demonstration of property of minute metallic particle due to quantized conduction band and the prediction of periodic frequency dependence of the electronic susceptibility for the set of particles of equal size and that changed the basic assumption for the energy level statistics [5]. Since then structures became an increasingly important derivative of their bulk due to their completely different optical, electrical, chemical and physical properties. Now a day it is well accepted that microstructures with characteristic length on the order of typically 1 - 100 nanometers are considered as nanostructures. The current (nanomaterials) NMs can be categorized into five material-based categories: (1) *carbon-based NMs*, which are composed mostly of carbon, such as carbon nanotubes (CNT), graphene nanosheets, and nanodiamond; (2) *inorganic-based NMs*, which include non-carbon NPs and NSMs such as metals (Au, Ag, and Cu NPs) and metal oxides (ZnO, TiO_2 and Fe_2O_3NPs); (3) *organic-based NMs* mostly containing organic matter, excluding carbon-based or inorganic-based NMs, such as nanostarch, nanocellulose, and polymer NPs; (4) *composite-based NMs* that combine NPs with other NPs or NPs combined with larger bulk-type materials (e.g., composite nanofibers) or more complicated structures, such as a metal organic frameworks and mixed metal oxides and (5) *bio-based* NMs, which are composed mostly of biomaterials such as enzymes and nanobacteria. The NMs can be also organized into a natural or man-made (engineered) based on their origin. The properties of nano materials largely rely on their structures, and thus many nanostructures have been synthesized to particular shapes, like particles, rods, belts, tetra- pods, spheres, owers, needles and tubular whiskers. These nano-shapes have good chemical and optical performance regarding their physio-chemical properties. High ratio of surface area to volume is an important characteristic for effective interactions between materials and foreign agents present in the proximity region. It is important to note that even if the number concentration is often dominated by nanoscale particles, their mass is usually negligible compared to that of coarse particles. Consequently, other metrics than mass may need to be taken into account in order to make an adequate and comprehensive evaluation. Unfortunately, it is not yet clear which key particulate parameters (mass, surface area, number concentration or their size distributions) could be the most relevant for any specific type of application.

They have potential for many chemical and biological applications, due to their large specific surface, high surface activity, strong antioxidant property, good biocompatibility, and suitableness for manipulations at the molecular level. At present, common nanomaterials used in biological applications include liposomes, polymeric micelles, graphene, carbon nanotubes, quantum dots, magnetic nanoparticles, metallic nanoparticles, and so on. And the nanostructures that are used for chemical sensing and application are gold, silver, Titanium, Carbon, Iron, Platinum. Their use has been shown remarkably enhanced sensitivity, selectivity and therapeutic outcomes. Amongst many advantages in using these nanostructures for any specific type of application for example. The chemical optical and electrical properties of nanostructures greatly depend on their shape and size. Secondly, due to the presence of electrically active surface charge on these nanostructures, they can be easily functionalized by all kinds of chemical and biological receptors, such as MPA, PDCA, Glutathione, drugs, genes and many other targeting ligands. Thirdly, most of these nanostructures are chemically and biologically compatible and most of them are nontoxic makes suitable. And moreover, they possess a distinct surface effect, ultra-small size, macroscopic quantum tunneling effect, and the presence of surface plasmon resonance (SPR) bands. All of these particular properties have caused nanostructures to become the most potential material for various biomedical and chemical applications, including chemical and bio-sensing, molecular imaging, drug carriers, and so on. Apart from all exceptional advantages of engineered nanostructures various disadvantages has also been reported such as inhaled nanomaterials have the potential to initiate inflammatory responses people inhaling are more likely to suffer from diseases of the cardiovascular system. There is the potential for nanomaterials to contact the skin and gastrointestinal tract as a result of workplace exposure. Recent research has identified that certain metal and metal oxide nanoparticles that are known to be inflammogenic also have a high positive zeta potential. This suggests that this parameter may be a useful predictor of certain types of toxicological effects. Some elements have been identified as carcinogens, mutagens, reproductive toxicants (collectively referred to as CMRs) or have the potential to cause asthma. It is reasonable to assume that a nanoparticle containing an element (s) that has these hazardous properties also has the potential to exhibit these hazards. The presence of reactive metals is thought to be responsible for the toxicity of complex particulate mixtures such as welding fume. A nanomaterial that contained a significant proportion of such metals (e.g., large amounts of catalyst residues in CNTs) could pose a greater hazard to health than a similar material with no or low amounts of reactive metals. For example, nano sized silver particles have a greater tendency to release silver ions into solution compared to larger silver particles. If dissolution leads to the release of reactive or cytotoxic components results adverse health effect. An early 2-year inhalation study of Degussa P-25 (a ~3:1 mixture of ~85-nm anatase and 25-nm rutile titania) resulted in lung tumors in rats [6]. SWCNTs containing residual catalytic metals produced greater pulmonary toxicity, including epithelioid granulomas and some interstitial inflammation, than ultrafine carbon black or quartz. These effects extended into the alveolar septa [7]. A review of eleven studies of carbon nanotube introduction to the lungs of mice, rats, and guinea pigs revealed most found granuloma, inflammation, and fibrosis. MWCNTs produced greater acute lung and systemic effects and were twice as likely to activate the immune system as SWCNTs, suggesting the former have greater toxic potential [8]. Engineered nano materials induced adverse health effects are reports that inhaled CNTs potentiate airway fibrosis in a murine model of asthma and that exposure of a cell line derived from normal human bronchial epithelial (BEAS-2B) cells to SWCNTs and graphite nanofibers produced genotoxicity and decreased cell viability [9, 10].

SYNTHESIS PROCEDURES FOR PLASMONIC NANOSTRUCTURES

Plasmonic nano-structures have unique properties in comparison to macroscopic ones. One of these is the possibility of the excitation of plasmon polaritons by light on their surface, which results in surface plasmon resonance (SPR) at certain excitation wavelengths. Surface plasmon polaritons are electromagnetic excitations propagating at the interface between a dielectric and a conductor, evanescently confined in the perpendicular direction. Photonic devices based on LSPR have been used to enhance the efficiency of fluorescence, Raman scattering, infrared absorption and even second harmonic generation. Therefore surrounding, matrix materials are important in such structures. The main challenge for LSPR active nanostructures is to develop efficient, sensitive, yet cost-effective plasmonic structures for various applications.

The ease of the preparation, customization, and reproducibility are important criteria for these nano-structures. However, by developing plasmonic nano-structures suitable LSPR active and fabrication of combined devices that can provide better sensitivity and a broader range of applications. Many efficient methods have been developed to achieve these goals some of them are following. (1) Chemical route of synthesis (2) Physical route of synthesis (3) Microorganism Assisted Synthesis (Scheme 1).

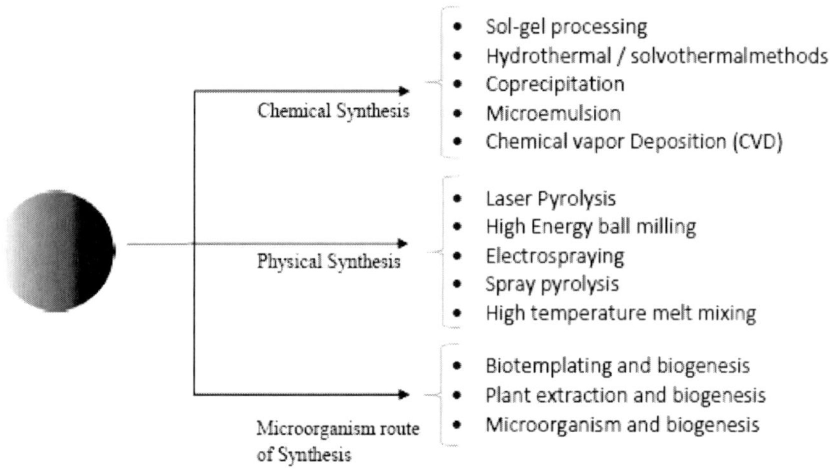

Scheme 1. Depiction of three major route for the synthesis of plasmonic nanostructures.

Physical route of synthesizing nanostructures mainly encompasses top-down strategy. The absence of solvent contamination during preparation and the uniformity of nanoparticle distribution is the biggest advantage of this route. Nanostructures synthesized by this method are usually highly stable. Many result has shown that geometric structure and total concentration of nanostructures can be greatly controlled without agglomeration according to the desire of the experiment. Some of the Physical method for example Laser pyrolysis technique is highly desirable because to produce nano-colloide with the highest purity by controlling the wavelength of laser impinging and the duration of the pulse starting from femto, pico and nanosecond, duration of ablation, effective liquid medium and with or without the surfactant. Mafune et al. [11] synthesized silver nanostructures by laser ablation in an aqueous solution by using ND-YAG laser at 532 nm operating at 10 HZ and 90 mJ/Pulse. Kabashin et al. [12] prepared Gold plasmonic nano structures using femtosecond laser ablated gold solution in water. In this article author reports a very small (3-10nm) and monodispersed gold colloid exhibits highly plasmonic properties and optical absorption peak which is potential for biosensing application. Biswas et al. [13] synthesize iron oxide nanoparticle in modified scaling law by electrosprying ionic precursor solution. The modified scaling laws are verified by comparing measured size distributions of the resultant electrosprayed aerosols generated by electrospraying aqueous and ethanol solution of ferric nitrate. They demonstrated the narrow size distribution, nanosized iron oxide particles produced by thermally decomposing the resultant electrosprayed iron nitrate precursor particles. Sol-gel process of making nanocomposites is an emerging as a very promising technique with enhanced nonlinear optical properties. The reasons for such a fruitful application of this process is due to molecular scale homogeneity of the starting solutions, the low processing temperatures, and the possibility of incorporating many different metal dopants into matrixes [14-17]. Biotempelating route for the fabrication of nanostructures is another extremely important and robust method to construct desired shape and sized nanostructure encompass interesting properties. Qi et al. [18] and coworkers synthesized Gold nanoparticle-Bacteria cellulose nanofiber, in this work Bacteria cellulose nanofiber were uniformly coated with gold nanoparticle in aqueous suspension with varying thickness which was successfully utilized for H_2O_2 detection as low as 1μM. Sastry et al. [19] demonstrated the use of spider silk for the one step synthesis of Gold nanostructure. They further demonstrate exposure of the gold nanoparticle-spider silk bioconjugate to vapors of methanol and chloroform leads to changes in electrical transport through the nanoparticles and thus, the possibility of developing a vapor sensor. Kim and coworkers [20] used Pine, Persimmon, Ginkgo, Magnolia and Platanus leaf extracts as a reducing agent for Ag^+ to Ag^0 followed by the formation of metallic silver nanostructure. They reported that Magnolia leaf broth acts as best reducing agent in terms of synthesis rate and 90% conversion to silver nanostructure within eleven minutes at 95^0C temperature. They also reported that Ag nanostructures synthesized by this route has an extraordinary inhibitory effect toward many bacterial strains and microorganisms commonly present in medical and industrial processes. A vast array of biological resources available in nature including plants and plant derived products for example algae, fungi, yeast, bacteria and viruses that could all be employed both on unicellular and multicellular organism to produce intracellular or extracellular inorganic nanostructures [21-29]. Wet chemical reduction is the most frequently applied method for the preparation of any kind of nanostructures as they are most stable colloidal dispersion in water or organic solvents. Commonly used reductants are borohydride, citrate, ascorbate and elemental hydrogen. Generally, reduction of metal ions in aqueous solution yields colloidal nanostructures with particle diameter of several nanometers. Initially, the reduction of various complexes with ions leads to the formation of respective atoms, which followed by agglomeration into oligomeric clusters. These clusters eventually lead to the formation of colloidal nanoparticles. Past couple of decades many authors including us has reported various sophisticated chemical design for the controlled synthesis of nanostructures by the potential hope for the applications and economic impact. Senapati et al. [30] has reported a detailed synthesis protocol for the evolution of branched gold nanostructures based on seeded growth approach. They also stablished the real-time monitoring of the shape evaluation during synthesis by controlling the ratio of encapsulating material. Misra et al. [31] used Ag^+ ions to catalyze the anisotropic growth of gold nanoparticle in the solution to make star shaped plasmonic nanostructure. Schultz and coworkers [32] has carried out an outstanding localized catalysis strategy on the surface group of a self-assembled monolayer with a scanning probe device to make nanostructure. In this work author used a platinum-coated atomic force microscope tip to catalytically hydrogenate terminal azide group with high spatial resolution. Given the large number of heterogeneous catalytic reactions involving transition metal catalysts, this approach describes a general chemical synthetic approach to engineer nano-scaled materials with desired surface properties by chemical conversion of chemical group and subsequent covalent attachment of molecule to these modified surface group for specific type of application. Controlled synthesis of plasmonic nanostructures on their size and tailored architecture are important for their specialized application. As an example polyhedral structures with the sharp corners and edges has significantly enhanced performance for catalytic, plasmonic, energy conversion and spectroscopic application [33, 34]. Chen et al. [35] has reported a fivefold Stellate Polyhedral gold nanostructure with [110] facets using seed mediate growth technique. Diameter dependent optical characterization revels that these fivefold stellate plyhedra morphologies arise from the deposition of Au on multiple twinned seeds. Multiple twinned structure Au seed prefer icosahedra growth with the total free energy lower than Wulffs polyhedron which derive to grate stellate dodecahedron. Huang and coworker [36] went one step further in advancing the synthesis interesting AuPd Nanowheels for the potential use for the efficient conversion of light into chemical energy. The key components of plasmonic material are noble metals, since the electromagnetic waves can be coupled to collective oscillations of free electrons in the metals. The plasmonic features of noble metal nanostructures are highly dependent on the size and morphology and composition. This synthetic strategy is based on the introduction of the oxidant HNO_3 which can mediate the reduction sequence of $HaAuCl_4$ and Na_2PdCl_4 that leads to the formation of unique AuPd nanowheels with only an Au core surrounded by metallic Pd. They have also shown that nanowheel found by route can be utilized solar energy to drive catalytic reaction such as the oxidation of benzyl alcohol and Suzuki coupling reactions at lower temperature and with much higher performance than possible with the conventional heating process.

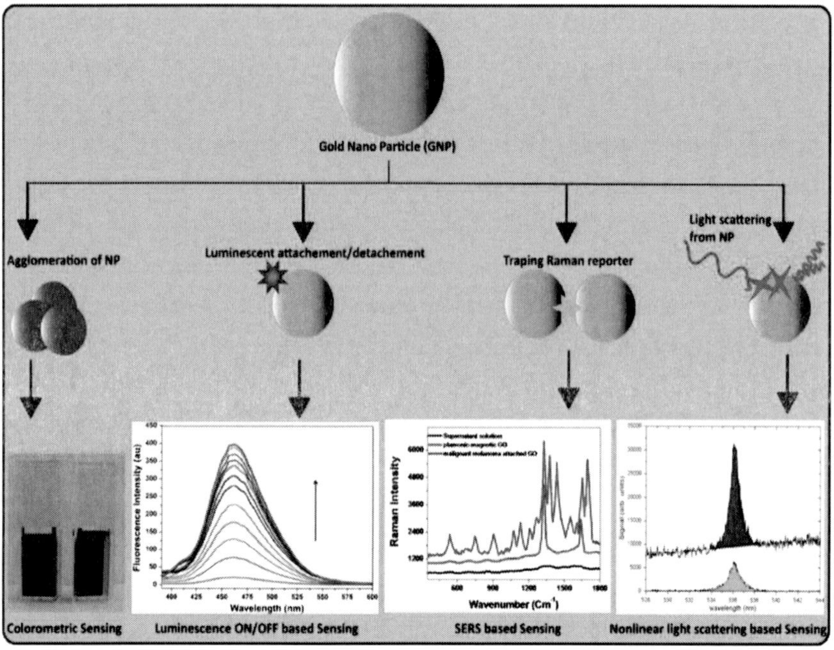

Scheme 2. Demonstration of four major spectroscopic tool to use plasmonic nanostructure as a probe for fabricating smart sensor.

Biological Application

Proven wide verity of application of engineered plasmonic nanostructure due to their unique physical and chemical properties, Recent investigations of functionalized nanomaterial have expended into the biological area and providing a versatile platform in biomedical applications such as biomolecular sensing, biological imaging, drug delivery and therapy. It is well proven that bio-functions and biocompatibility of nanostructures can be realized by introducing synthetic ligands or natural biomolecules onto nanomaterials surface and combining ligand receptor biological interactions with intrinsic nanomaterial and properties. Common strategies of engineering nanomaterial surface involve physisorption or chemisorption of desired ligand. Relying on intrinsic nanomaterial properties in conjunction with immunogenic or other biological recognition mechanism, detection schemes colorometric, fluorescence turn ON/OFF, surface enhanced Raman scattering (SERS) and nonlinear optical techniques are broadly exploited mechanisms for bio agent threat detection. A schematic representation the mechanism is depicted in the (Scheme 2). Many of techniques highlighted are research based prototypes not yet ready for integration into commercial device, but they already offer promise when compared to the current gold standards for detection and highlight areas in which nanosensors are able to outperform the traditional detection paradigms. Here we provide an overview of some popular nano based approaches illustrate them with current examples showing how they have been applied and discuss their intrinsic benefits and potential labilities.

Colorimetric Based Sensing of Biological Pathogens

Early and accurate detection of biological pathogens are key to prevent long time impact on health care in today's word. Accurate detection is often time-consuming techniques and expansive instrumentation. To address these limitations many group including us has been involved in designing colorimetric assay for the early direct diagnosis of the diseased cells. Normally colorimetric assays available in the market uses antibody and aptamer conjugated nanostructures to combine the selectivity and affinity of conjugated nanostructures and the spectroscopic advantages of these nanostructures to allow for the sensitive detection of pathogens. In this technique samples with the target cells exhibits a distinct color change while nontarget samples do not elicit any change in color with both naked eye and based absorbance measurement. In many cases these assays differentiate type of target and control cell based on the antibody, aptamer used in the assay indicating the wide applicability for diseased cell detection. Tan [37] and coworkers has developed an assay that is colorimetric in in nature and has excellent selectivity between target and control cells Figure 1.

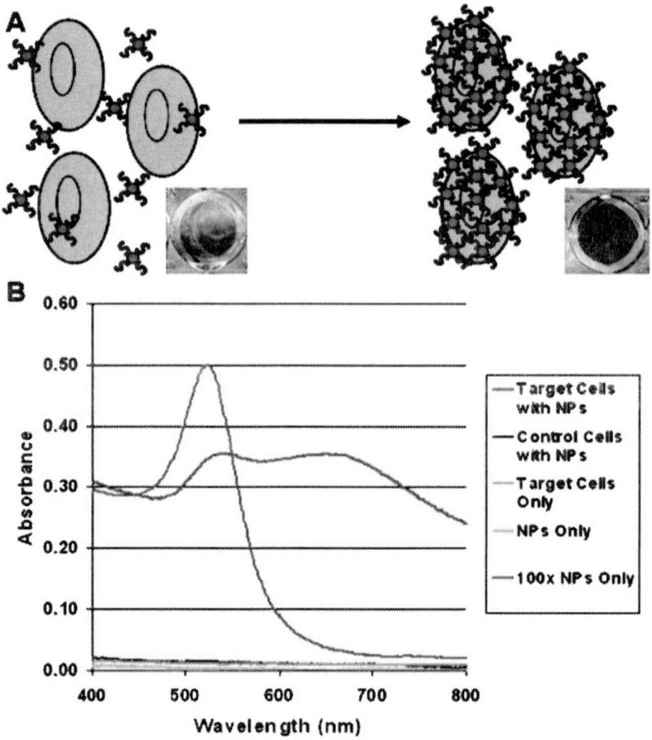

Figure 1. (A) Schematic representation of aptamer conjugated gold nanoparticle based colorimetric assay. (B) Absorption spectra of various samples analyzed using aptamer conjugated gold nanoparticles. (Adapted with permission from *Anal. Chem.*, 2008, *80* (4), pp 1067–1072. Copyright, 2008 American Chemical Society)

In this work, to make this assay colorimetric in nature, surface of the gold nanoparticles were modified for their response to target cells with respect to spectral properties. Once functionalized AuNP come into proximity with one another their absorption spectra shift and their scattering profile changes due to the change in dielectric and plasmon properties resulting a change of color of the colloidal solution. This technique has also widely used to track chemical toxins, proteins and genes [38-43]. Surface functionalization with appropriate biomolecule such as aptamer, DNA, peptide or antibody is the most important to use this technique. These biological functional group can be stabilized either noncovalent or through covalent interactions. Noncovalent stabilization of functional group at nano-surface takes place through electrostatic interactions, hydrophobic entrapment, and van der walls force. The advantage of this interaction is that the functional group cannot attached to various chemical modifications that are likely to compromise its primary active modality. Covalent functionalization of nano-surface is the use of immediate chemical attachment, linker molecules or click chemistry provide a greater stability and reproducibility. This kind of modification can withstand a very high salt concentration and are extremely stable under thermal conditions. Moreover, covalent functionalization are normally are more complex then noncovalent functionalization and this requires intensive synthesis work on functional ligands.

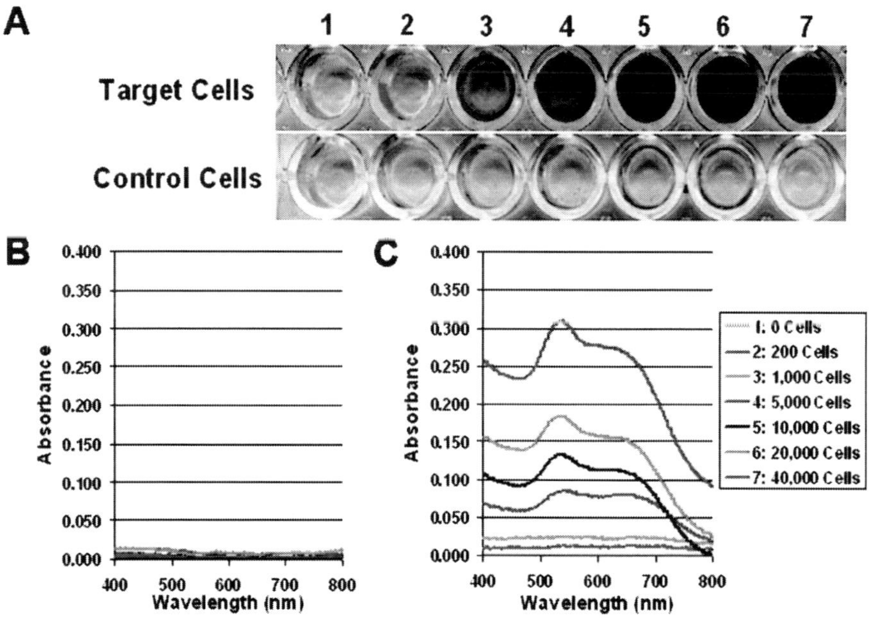

Figure 2. (A) Photograph the visual color change after changing the concentration (1 – 7) of target cell and control mixed with aptamer conjugated plasmonic gold nanoparticle. The amount of cells used in each sample is given in the legend on the bottom right. (B) Absorption spectra of the control cell samples with aptamer conjugated plasmonic gold nanoparticle. (C) Extinction spectra of the aptamer conjugated plasmonic gold nanoparticle with varying concentration of target cancer cells. (Adapted with permission from *Anal. Chem.*, 2008, *80* (4), pp 1067–1072. Copyright, 2008 American Chemical Society)

In this particular work to demonstrate the principle behind the colorimetric assay, an aptamer sequence with high selectivity and affinity to cancerous acute leukemia cell was covalently conjugated to 20 nm gold nanoparticle through thiol functional group on the aptamer sequence (Figure 1).

The excellent selectivity of the assay is achieved due to highly selective aptamer utilized in the experiment and also from the nature of the gold nanoparticles themselves. Sift in the absorption profile before and after interaction with target cell can be visually differentiated on the absorption spectrum. In this case, after binding of the target cell with bio conjugated gold nanostructure now behave as a larger gold structure, the light scattering and absorption properties of the nanostructure are altered as well resulting in the majority of the signal transduction in the assay. Due to change in the optical properties followed by visual change in the color the sample possibly allows to design a more sophisticated and even simple naked eye analysis methods to use these plasmonic nanostructures for biological molecules. As shown in Figure 1 (B), there is significant change in the absorption spectra with the amount of target cell. In other way sample absorbs more light when the target cell added in the bio functionalized nanocolloid solution. In this work author has also shown the analysis of target sample using microplate reader where the color of the sample can be visually differentiated when the target sample is mixed with nanoparticle solution. The variation in the color of the solution in the microplate reader correlate exceptionally with the recorded absorption spectra of the mixture. Findings of this experiment stablishes that the aptamer conjugated gold nanoparticle are binding selectively to the target cell and that increases absorption and scattering properties of the solution. Since this technique rely on the large scale assembly of the particle to change their spectral properties the nonspecific binding of a small amount of nanoparticle could cause the generation of a false positive signal. Based on the type of target cell, many types of surface modification methods have been developed by many authors [44-47]. Luo and co-workers [48] has used DNA to assemble gold nanoparticle to selective target in vA gene of salmonella bacteria. They used a DNA bind plate to immobilize gold nanostructure to form a firm probe. This sensing platform not only enlarge the target library of the DNAzyme probes based gold nanoparticle colorimetric sensors but also provide a new simple and economical method to detect salmonella and various other type of biological targets. In our previous work, we have also demonstrated this nanotechnology driven approach to selectively target and destroy pathogenic bacteria [49]. In our experiment we have demonstrated the possible use of anti-salmonella antibody to functionalize covalently on the surface of oval shape gold nanoparticle to track Salmonella typhimurium with an excellent discrimination (10^4 bacteria per mL) over other pathogens. In order to evaluate the sensitivity of this antibody modification method, different concentration of Salmonella tryphimurium bacteria from one stock solution were evaluated and demonstrates that the sensitivity of this colorimetric assay is extremely high and comparable with the detection limit of enzyme-linked immunosorbent assay (10^4 – 10^5cfu/mL) and antibody-microarray biochip technique. Moreover, it is well known that the size and shape of the gold nanoparticles can be tuned based on the synthesis protocol to strongly absorb NIR radiation that can generate heat and that further damages the wall of the attached target cell. Using this technique, we have further demonstrated that when the oval shaped gold nanoparticle are attached to the bacteria cells, the localized heating that occurs during NIR radiation causes irreparable cellular damage. Our experimental result therefore clearly demonstrates this bio conjugated colorimetric assay not only selectively track salmonella but could be the next avenue for exploration towards selectively targeting and destroying microbial pathogens.

Luminescence ON/OFF Based Microbial Sensing

It is well demonstrated that nanostructures have very high molar extinction coefficient and can be treated as non-molecular chromophore and has exceptionally high quenching ability [50-52]. Many authors including us has successfully utilized these extraordinary abilities of nanostructures to construct energy acceptor FRET assay for biological and chemical sensing. Rohadamin 6G, Rohadamin B, dyes are well known fluorophore whose emission can be effectively quenched by gold nanoparticle [53-56]. This is an electrostatic driven approach in which attached dye molecule on the surface of gold nanoparticle can be dissociated to undo fluorescence quenching by addition of electrolyte such as sodium chloride. Moreover, by optimizing gold nanoparticle surface functionality and masking agents in the sample solution the selectivity via this technique can be can be over ~ 50-fold over other interfering pathogens. In comparison with common molecular quenchers, gold nanoparticle based quenching mechanism exhibits a sensitivity enhanced up to 100-fold with significantly discriminating over other interfering pathogens. Very recently we developed a methylene blue bound of plasmonic gold nanoparticle capable of sensing human prostate cancer cell LNCaP selectively using fluorescence imaging [57]. Since methylene blue is well known light active photodynamic therapy drug, in this work we have also demonstrated that using gold nanoparticle modified with methylene blue therapeutic action can be controlled with external NIR light. Transmission electron microscope image In Figure 3 (A), clearly shows the formation of gold nanostructure with small spike indicating roughened surface, energy dispersion X-Ray image in Figure 3 (B) demonstrate the presence iron core and gold, and absorption spectrum in Figure 4 (C) confirms the plasmonic nature of the nanostructure with absorption peak maximum at 780 nm. In order to confirm the possible use of nanoplatform developed by us for combined synergistic therapeutic action for cancer, we have used the LNCaP human prostate cancer cell line, which is well-known to overexpresses the prostate specific membrane antigen (PSMA) on the cell surface. For targeted imaging and synergistic therapy, we modified the nanoplatform with methylene (MB)-bound A9 RNA aptamer, which is known to be specific to PSMA.

For this purpose, initially, the core–shell nanoparticle was coated with thiolated polyethylene glycol (HS-PEG), which helps to avoid nonspecific interactions with cells and cell media.

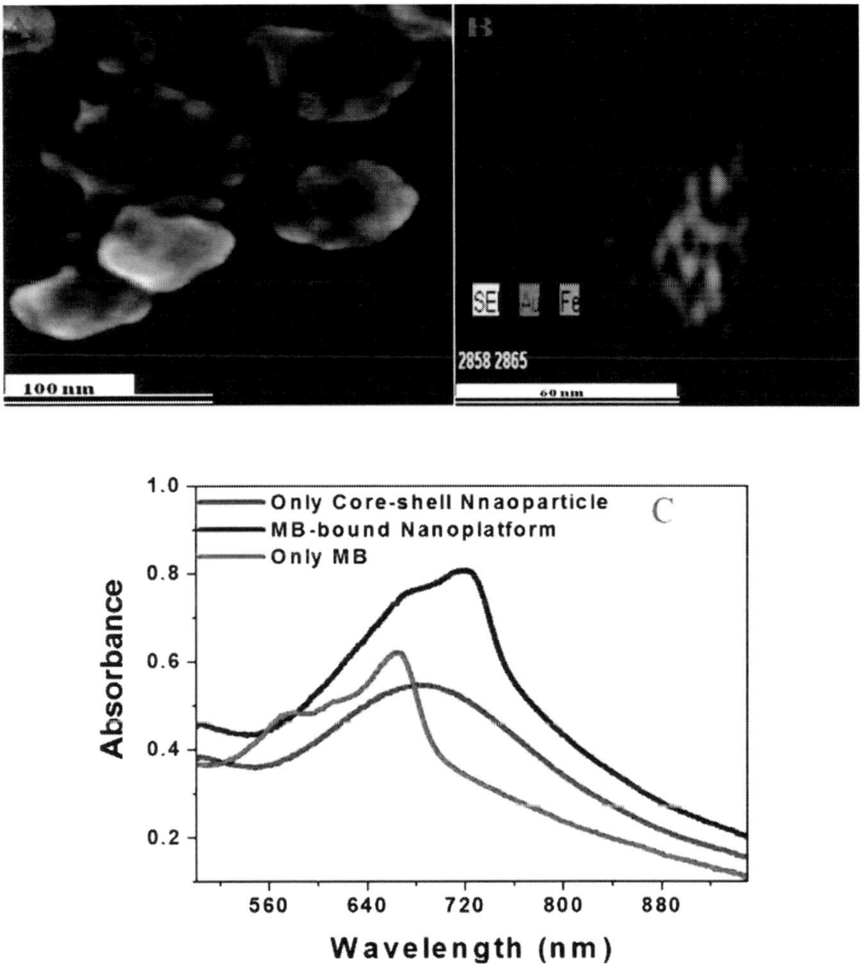

Figure 3. (A) Scanning electron microscope image of core-shell plasmonic nano-platform. (B) EDX mapping image shows the presence of iron (core) and gold (outer plasmonic layer) in a core-shell nanostructure. (C) Absorption spectra of core-shell nanoparticle, methylene blue, and methylene blue conjugated nano-platform. The strong long wavelength band in the near IR region for the nano-platform is due to the oscillation of the conduction band electron of gold. (Adapted with permission from *Mol. Pharmaceutics*, 2014, *11* (4), pp 1109–1116. Copyright, 2014 American Chemical Society)

After PEGylation, thiol-modified MB-bound A9 aptamers were attached via thiol chemistry technique. Next to show the capability of our nanoplatform to separate cancer cells from normal cells, LNCaP cancer cell suspension containing 10^3 cells/mL was mixed with 1 mL of PSMA-negative human skin HaCaT cell suspension containing 10^7 cells/mL. After that we added 100 μL of nanoplatform colloidal solution to the cell mixture and incubated for 120 min under gentle shaking. Next, to separate nanoplatform attached cancer cells from the mixture, we used a bar magnet. At the end, after magnetic separation, we used fluorescence image, TEM, and enzyme-linked immunosorbent assay kits for the characterization of cells that bind with nanoplatform and separated them by bar magnet. Similarly, cells that did not bind with the nanoplatform were also characterized using microscopic and optical imaging technique as well as using enzyme- linked immunosorbent assay. Our enzyme-linked immunosorbent assay results indicate no PSMA presence in the fractions of cell suspensions that did not bind to the nanoplatform, which indicates the absence of LNCaP cells in cell suspension which are not separated by the magnet. On the other hand, we find that PSMA was present in the nanoplatform attached cell suspension. Enzyme-linked immunosorbent assays results clearly indicate that LNCaP cells are attached to the nanoplatform. From the enzyme-linked immunosorbent assay experiments we have estimated that the LNCaP cancer cell recovery by bar magnet was about 97%.

Fluorescence imaging data using confocal fluorescence micro- scope, as shown in Figure 4 A−D, clearly shows that the nanoplatform was bound only to LNCaP cells and not HaCaT cells. Similarly, TEM images (Figure 4 E, F) indicate that the nanoplatform attached only to LNCaP cells and not normal HaCaT cells. All the experimental results, described above, clearly indicate that the MB-bound A9-aptamer-conjugated nanoplatform developed by us is highly selective for binding with the LNCaP prostate cancer cell line, which overexpresses PSMA. Next, to find out whether our design of core−shell nanoplatform can be used for multimodal therapy, after successful targeted prostate cancer LNCaP cell separation, we performed NIR irradiation experiments using 785 nm excitation light. For this purpose, at first, we performed a cytotoxicity experiment in the absence of NIR light to determine whether the nanoplatform developed by us is cytotoxic or not. Next, to determine if the combined therapy is much superior to single therapy, we designed multi step experiments. For the cytotoxicity measurement, MB-bound nanoplatform attached LNCaP cells were incubated for 12 h without any light exposure. We did not observe any cell feature even after 12 h of incubation, which indicates that the nanoplatform developed by us is not cytotoxic in the absence of external NIR light. This fluorescence imaging technique has been explored since a decade and proven to have potential to identify non-advanced, even pre-cancerous tumor where other imaging based on white light or radiation overlook them. Due to strong electric fields at the surface of plasmonic nanostructures enables the possibility of manipulation of surface properties that provides the potential of designing novel optically active agent for simultaneous molecular imaging and photothermal therapy. El-Sayed and his group [58] in their very pioneering work designed a gold nanorod with aspect ratio that can absorb and scatter strongly in 650 – 900 nm region which is suitable to minimize the light extinction by intrinsic chromophore in native cancerous tissue. They used this specially synthesized gold nano-rod with specific length and diameter (aspect ratio) as a contrast agent for both molecular imaging and selective photothermal cancer therapy. In this work they functionalized the surface of the plasmonic nanorod with anti-epidermal growth factor receptor (anti-EGFR) to selectively target desired cancer cell. This special rod like structure of nano-gold exhibits two plasmon bands one at around 520 nm due to transverse electronic oscillation and another in NIR region (depending on aspect ratio) due to longitudinal electronic oscillation. Selection of specific aspect ratio and radiation wavelength being used for photothermal therapy is important due to their absorption overlapping with a region of minimum extinction of human tissue. This conjugated gold nanorod with anti-EGFR monoclonal antibody specifically targets the molecular marker EGFR to identify them selectively and acts as photothermal absorbers for destroying cancer cell with NIR laser radiation without affecting the surrounding nonmalignant cells.

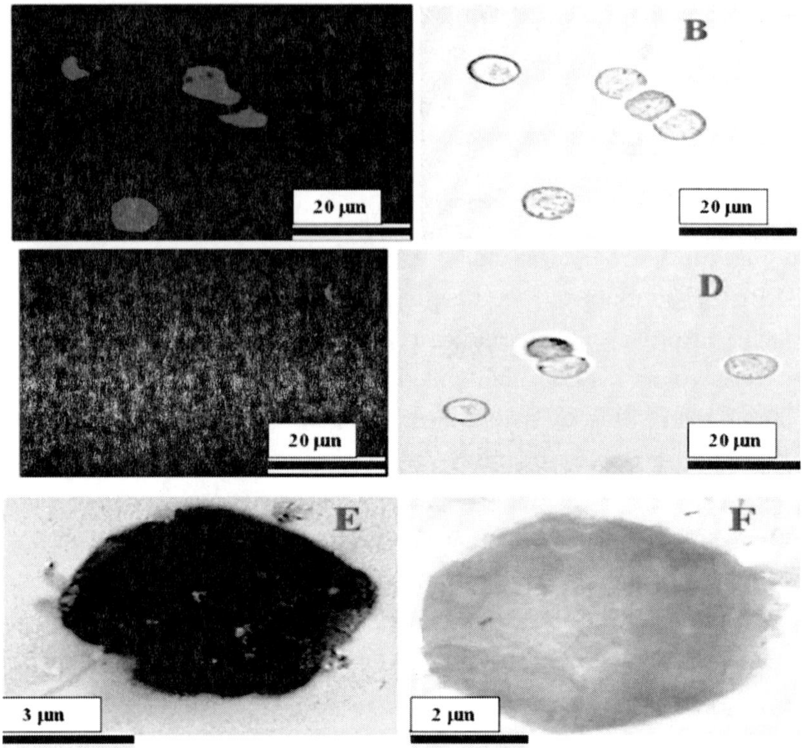

Figure 4. (A) Fluorescence microscope image of nano-platform attached LNCaP human prostate cancer cell, after LNCaP cells were separated from cell mixture using a magnet. (B) Bright field microscope image of the same LNCaP cells after magnetic separation. (C) Fluorescence image of cell suspension which are not separated by magnet. (D) Bright field microscope image of cell suspension which are not separated by magnet. (E) Transmission electron microscope image (TEM) of plasmonic nano-platform sample attached with LNCaP cells after magnetic separation. (F) TEM image of suspension of cells which are separated by the magnet, which is mostly skin cells. (HaCaT cells.) (Adapted with permission from *Mol. Pharmaceutics*, 2014, *11* (4), pp 1109–1116 Copyright, 2014 American Chemical Society)

SERS Based Sensing of Bio-Pathogens

Surface-enhanced Raman scattering (SERS) technology is an ultrasensitive vibrational spectroscopic technique for measuring signature vibrational spectra of Raman active molecules. When these molecules are in near proximity of the surface plasmon of nanostructures, their Raman scattering profile can be increased by many fold. SERS techniques has been applied in biological and environmental contaminant sensing for many years. The measure advantage of this technology is its excellent photo stability due to need of ultra-low dose of laser radiation to observe scattering signal. Recently this technique has recently used by many researchers including us to design noval nanoprobes that encapsulate Raman reporter molecules in the hot SERS active sites of plasmonic nanostructures [59-63]. Moreover, placement of these SERS active tags near the bio-pathogen enables to track biomolecular activity, conformational changes, understanding drug delivery mechanism and cell biology. Recently we designed a distance dependent SERS assay for monitoring photothermal antibacterial activates.

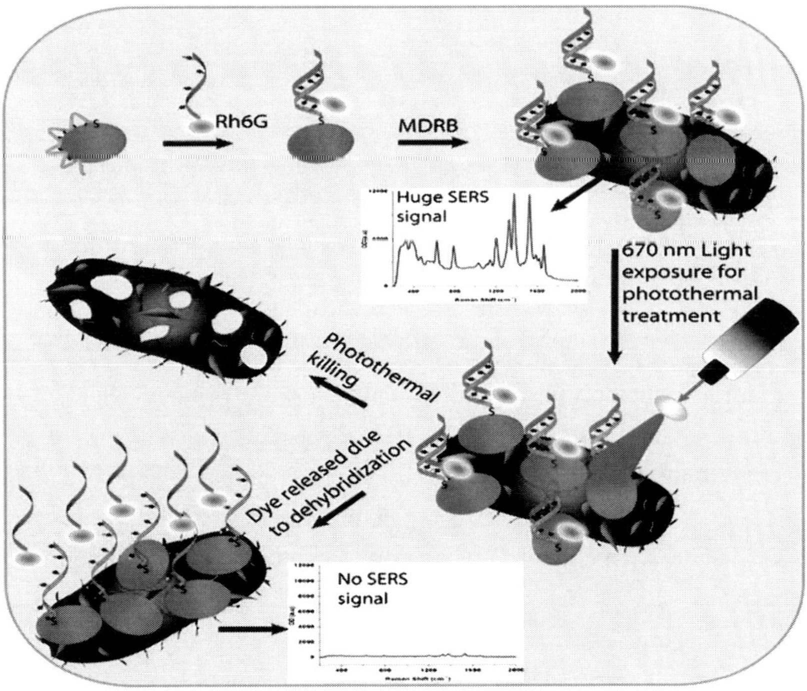

Figure 5. Scheme shows use functionalized plasmonic nanostructure for the development of the distance dependent SERS assay for monitoring photothermal bacterial killing activity. (*Chem. Commun.,* 2012, 48, 11091–11093)

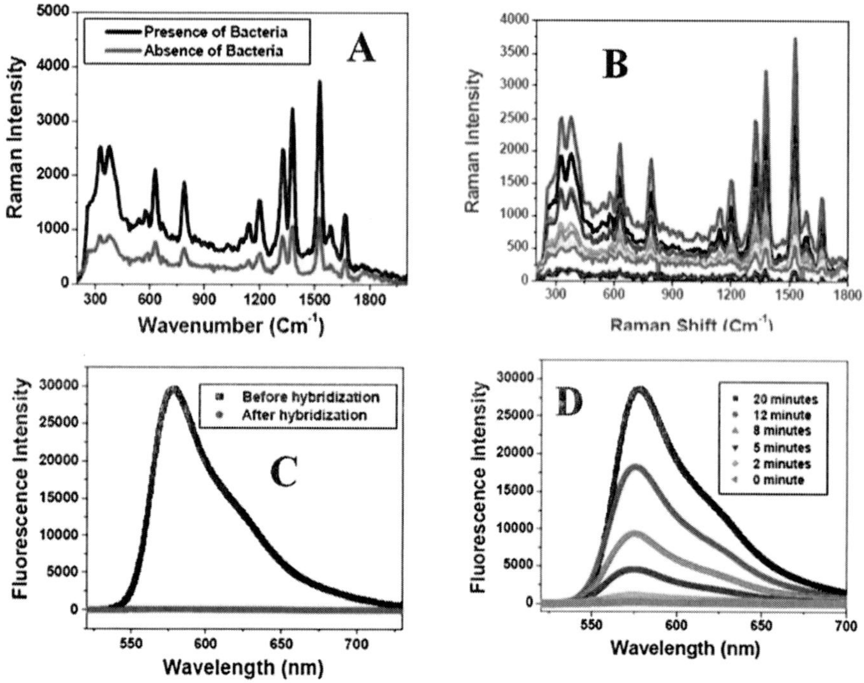

Figure 6. (A) Spectroscopic response graph shows a significant SERS intensity enhancement after the addition of 1.3 x 10^5cfu/mL MRSA. (B) Graph demonstrates the time dependent SERS intensity decrease during the photothermal antibacterial activity process. (C) Plot shows how fluorescence intensity varies when Rh6G modified ss-DNA is hybridized with extended aptamer attached oval shaped plasmonic gold nanostructure. (D) Plot demonstrates time dependent fluorescence intensity increasing during the photothermal process. (*Chem. Commun.*, 2012, 48, 11091–11093)

Figure 5, depicts the technology and working principle of this SERS assay for tracking photothermal bacteria destroying mechanism. To do the objective in this case, we synthesized and used bio conjugated oval shaped plasmonic gold nanostructure to target the bacteria. More specifically, to design distance dependent SERS assay for monitoring the photothermal antibacterial activity process, oval shaped gold nanoparticles were coated with double-stranded DNA, where one strand of the dsDNA possesses a terminal thiol moiety on its 3'-end for the attachment to the nanoparticle surface. As shown in Figure 5, in this configuration, the thiolated strand of the ssDNA serves as the host molecule. The complementary DNA modified with Rh-6G hybridized to the host molecule is nonthiolated. As a result, after hybridization, Rh-6G modified complementary DNA was bound only to its DNA complement host and not to the nanoparticle surface. Since under these conditions, Rh-6G is placed very near to the surface, we have observed a very high SERS signal. Upon 670 nm laser illumination, during photothermal lysis, the lattice heat content is sufficient enough to raise the local temperature above the melting temperature of the DNA duplex. As a result, the dsDNA was dehybridized and will release the Rh-6G modified non-thiolated ssDNA sequence. As a result, the distance between oval shaped gold nanoparticles and Rh-6G Raman dye changes abruptly SERS turn off, and our experimental data show that the SERS intensity decreases by a huge amount [64]. For the selective imaging and killing of MRSA, we have used aptamer APT^{SEB1} extended sequences. Recently it has been reported that aptamer APT^{SEB1} can be used for very selective detection of staphylococcal enterotoxin B [65, 66]. In the transmission electron microscope image (TEM) we have noticed the clustering of nanostructures on the MRSA cell surface that opens the possibility of Raman dye to trap close enough to surface plasmon of nanoparticle. In order to identify whether these accumulations of nanostructure near the bio surface is specific for MRSA or obvious for any biological agent, we treated the bio functionalized gold colloid with E. Coli, we did not notice any significant labeling of nanostructures on the E. Coli membrane surface which indicates, since nanoparticle aggregates on the surface of the MRSA bacteria results in the formation of Raman active hot spot results in SERS enhancement from rhodamine 6G dye. As reported by several research groups, the largest Raman scattering enhancement are observed when molecules reside in the fractal space between aggregated colloidal nanoparticles [67-69]. As discussed in pervious sections these special nanostructures could be useful to generate localized heat to damage the sites in the close proximity. In this work, we have tried to monitor the possibility to use oval shaped nanostructure to create photothermal damage to microbial MRSA. Since the oval shape nanoparticle we use in this experiment have very high absorption at 670 nm light, we assumed that oval shaped gold nanoparticle will absorb this wavelength and convert optical energy into thermal energy disrupting the

MRSA cell membrane. This process may also help in optically eliciting the controlled release of attached functional group at gold nanoparticle surface help to monitor the activity near nano-bio interface. Figure 6 shows the result Raman and fluorescence spectrum of Rh 6G dye attached on the gold surface at various conditions. Since dye is attached with single stand of the DNA, in order to dye release the dye form gold surface, DNA duplex has to denature or Au-S bond has to be broken. To confirm, at first, we measured the DNA melting curves by monitoring the fluorescence in the gel column as a function of solution-ambient temperature without gold surface and found 36^0C at 40 mM salt concentration.

Thermal imaging of gold nanoparticle attached MRSA when exposed to 670 nm laser light revels that the local temperature raised above the melting temperature (~38^0C) to about 49^0C. This will trigger the dye modified strand to dissociate into the surrounding medium while its complement still attached to the nanoparticle as shown in Figure 6. Under these conditions, a marginal change in the distance between oval shaped gold nanoparticles and Raman active dye results in the SERS intensity as shown in Figure 6 (A& B). Result in this work also demonstrate that simply SERS intensity change due to duplex DNA melting can be used to monitor *In situ* photothermal antibacterial activity. Since the Rh6G is also known for its high fluorescence, we also performed the distance dependent fluorescence experiment during the photothermal process. As plasmonic gold nanostructures are very good quencher, when Rh6G dye is very close to oval shape nanoparticle, most of the fluorescence emission will be quenched by gold nanoparticle as shown in Figure 6 (C& D). During 670 nm light exposure process temperature starts rising and reaching optimum value to melt DNA results the increase in the distance between dye and surface plasmon of nanostructure as a result quenching efficiency decreases and fluorescence intensity increases which can be clearly seen in Figure 6 (C). Figure 6 (D) also indicates progress of oligonucleotide release process in terms of fluorescence intensity change when exposed with 670 nm laser light. Similar type of work has been done by many research groups, Emory and coworkers [70] has used Rh6G to adsorb on Silver nano particle to probe them individually from a large heterogeneous population. They proposed for the very first time that intrinsic Raman enhancement factor can be amplified of the order of 10^{14} to 10^{15} for Rh6G molecule adsorbed on the nanoparticle surface. Sharp fingerprint like pattern obtained in SERS method for biological studies is the key advantages over fluorescence technique. SERS patterns are extremely distinct from other interferences within the complex biological media. Whereas fluorescence spectra could be disguised by strong scattering signals from cells and protein clusters at low signal intensities. The distinctive Surface enhanced Raman scattering pattern also eliminates the need for tedious separation procedure for the biological samples. Studying the dynamics and evolution of biomarkers at the biological cell and nanostructure interface is could greatly contribute to the understanding pathogens and/or cancer cell proliferation, thus identifying the SERS profile of different Raman tags could provide key information for the development of SERS based targeted therapy where conventional fluorescence based labels suffer from photobleaching and spectral crosstalk making them unsuitable for many applications.

NONLINEAR OPTICS BASED TECHNIQUE FOR NANO-BIOSENSING

Biological samples are sometime very rare for example circulating tumor cells (CTCs) and the amount available in sample is very limited, such analysis presents formidable analytical and technical challenges. Various techniques including FDA approved Cytological assays system utilizes positive selection of antibodies coupled on ferro fluids enrichment steps but due to high heterogeneous expression of cancer cells this technique faces a measure disadvantage on specific antibody positivity. Overall limit of detection in case of CTC is not limited by enrichment steps or addition of extra CTC identifiers but also the improvement in instrumental methods to examine the presence of CTCs. This limitation must be taken into account before any analysis is started, especially in this case, where subpopulations of metastatic tumor cells often do not express enough EpCAM to detect CTCs in early stage disease progression. Multiphoton microscopy coupled with nonlinear optics has opened up the possibilities of new discovery for breakthrough in biological sensing. When an intense light of frequency ω, exposed on a matter, it can generate a spectrum of the scattered radiation that includes bands with frequency 2ω, due to Two Photon Scattering. This technique can be easily applied to study a very wide range of materials because electrostatic fields and phase matching are not required. The high scattering ability of plasmonic nano particles has opened various ways toward different scattering based sensing methods. Interestingly two photon scattering properties of plasmonic nanostructures exhibits significant enhancement when they are aggregated, as compared to the individual components, due to their interparticle plasmon coupling. Therefore, a clever design of chemical interactions between plasmonic nanoparticle surface and biological species that would result in aggregation can be optically probed by simple measuring scattering signal with respect to biological species concentration. In the last decades, our group and other has demonstrated that due the presence of high surface plasmon resonance, week nonlinear effect can be significantly enhanced by about 4 – 6 order of magnitude using strong electromagnetic magnetic field at the nanosurface and that can be used as a fingerprint for the identification of CTCs after capturing and isolation from blood samples [71]. Several reports from our group and others have shown that TPS gives a sharp fingerprint-like spectral pattern which can be used to distinctly identify a particular bio-molecule from other interferences within the complex biological sample. Since TPS has been used to discriminate cell cycle phases and separate live from dead cells, gives opportunity to develop a versatile bio-conjugated gold nano platforms to detect and isolate CTCs from blood samples, and as a facile TPS probe for accurate identification amongst different types of CTCs. Over past decades several groups are engaged in exploring the possibilities to utilize highly sensitive NLO properties of nanostructures as probe for the selective identification of biological pathogens [72-74].

In one of our previous work we demonstrated that TPS properties of oval shaped gold nanoparticles can be used for rapid, highly sensitivity detection of human breast cancer cell line. This TPS of hyper-Rayleigh scattering based approach is based on the fact that bio-conjugated multifunctional oval shaped gold nanoparticle can readily and selectively identify the breast cancer cell line through HER2/c-erb-2/Neu receptor recognition as shown in Figure 7. For the SK- BR-3 cell line, there are many surface epithermal growth factor receptors (HER2/c-erb-2/Neu) available for selective recognition with monoclonal anti-HER2/c-erb-2 antibody and S6 aptamer conjugated ovals shaped gold nanoparticle. Therefore, in the presence of the SK-BR-3 cell line, several nanoparticles can bind to HER2 receptors in the cancer cell, thereby producing nanoparticle aggregates. We have clearly observed the binding multifunctional oval shaped nanostructure on breast cancer cell surface through the transmission electron microscopy and thereby formation of massive nanoparticle aggregates resulting a clear-cut alternation of optical characteristics of nanostructure in terms of visual change in color of the solution and thus absorption spectra of specimen colloidal solution. This change in optical properties is attributed due to two factors, one is the change in the local refractive index on the nanoparticle surface caused by the specific binding of the conjugated nanoparticle which binds to HER2/c-erb-2 on the cancer cell surface. Other is the inter-particle interaction resulting from the assembly of nanoparticles on the cell surface. We did observe some extremely poor labeling in the TEM image of for example healthy cell, skin cell and other type cancer cell with nanostructures may be due to nonspecific interaction between the antibodies and the proteins on the cell surface and thus the nanoparticles are randomly distributed on the whole cells. These results clearly demonstrate that this specifically conjugated nanostructure is highly selective for SKBR3 cell lines only and it is able to distinguish between healthy and different cancer cell line in the assay up to 10^4 cells/mL. Next to examine the use of two photon scattering as an alternative to improve the assay sensitivity towards the breast cancer cell (SKBR3) detection, we mixed various concentration of monoclonal anti-HER2/c-erb-2 antibody and S6 aptamer coted oval shaped gold nanoparticle with various concentration of breast cancer cell line. After that we used our own reported experimental setup to measure two photon scattering signal from the cancer cell and gold nanoparticle solution [71]. In brief, for TPS or hyper-Rayleigh scattering (HRS) experiment, we used a mode Ti-sapphire laser delivering a fundamental wavelength of 860 nm with pulse duration of about 150 fs at repetition rate of 860 MHz. We performed TEM data analysis before and after exposure to the laser beam to confirm any existent photothermal damage due to laser exposure during TPS data collecting time. The hyper-Rayleigh scattering signal was separated from its linear counterpart by a high-pass filter, 3 nm band-width interference filter and a monochromator and then detected it with a cooled photomultiplier tube. As shown in Figure 8 we see a distinctive change in the polar plot of two photon intensity as a function of the incoming beam polarization angle even upon the addition of as low as 100 SKBR3 cancer cell/mL where as in colorimetric experiment we observed color change only after addition of 10^4 cell/mL.

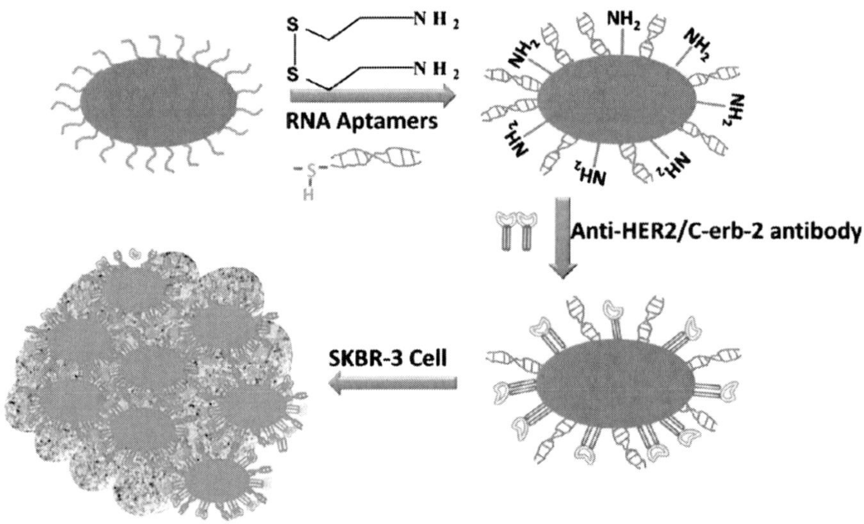

Figure 7. Step wise Schematic demonstration of the synthesis of monoclonal anti-HER2 anti body and S6 RNA aptamer conjugated oval shaped plasmonic gold nanostructure and the sensing technique of breast cancer cell line. (Adapted with permission from *ACS Nano*, 2010, *4* (3), pp 1739–1749. Copyright 2010 American Chemical Society)

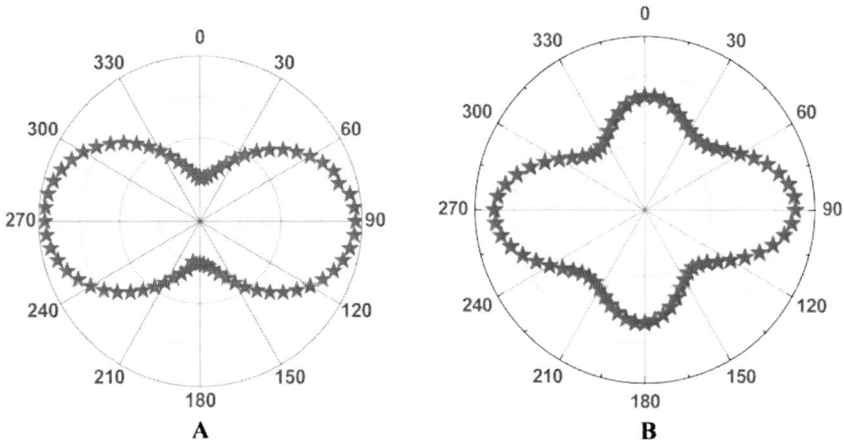

Figure 8. Two photon scattering signal plot as a function the incoming fundamental beam polarization angle from the conjugated oval shaped plasmonic gold nanoparticle. (A) Before addition of cancer cell and (B) after the addition of cancer cells. (Adapted with permission from *ACS Nano*, 2010, *4* (3), pp 1739–1749. Copyright 2010 American Chemical Society)

This result shows that hyper-Rayleigh scattering assay method is about two orders of magnitude more sensitive than the normal colorimetric assay. As far as two photon scattering is concern, this is a powerful method to determine the small change in the size of the particles and therefore even has capability to separate dimer from the monomers. Considering the size of the nanoparticle, the approximation that assumes that the electro-magnetic fields are separately constant over the volume of the particle may not be suitable anymore. As a result, the total nonlinear polarization consists of different contribution in the output intensity pattern of the polar graph such as multipolar radiation of the harmonic energy of the excited dipole and possibly of higher multipoles, as discussed in our previous publications and reported by other workers [75, 76]. The intensity change upon aggregation of conjugated nanostructures in presence of SKBR3 cell line consists of many contributions.

The very first and obvious is the electric dipole approximation, which may arise due to the defects in the nanoparticle. This is identical to the one observed for any non-centrosymmetrical point-like objects. After the aggregation in the presence of SKBR3 cells, conjugated gold nanoparticles lose the center of symmetry, and as a result, one can expect a significant amount of electric dipole contribution to the two-photon scattering intensity. Since the electric dipole contributes several times higher than that of multipolar moments, results two-photon scattering intensity to increase upon the addition of SKBR3 cells. The second contribution is multipolar contribution like electric quadrupole contribution. This contribution is very important to take into consideration when aggregation occurs due to the addition of SKBR3 cells. Since after aggregation the size of the particle is no longer negligible compared to the incoming wavelength and no longer negligible. As a result, after aggregation one can expect very high change in the intensity of polar plot of collected scattering signal. To probe this multipolar pattern, we performed angle-resolved HRS measurement. For this purpose, the fundamental input beam was linearly polarized, and the input angle of polarization was selected with a rotating half-wave plate. As shown in Figure 8, two lobes for multifunctional oval-shaped gold nanoparticles, which are similar to the one for pure electric dipole response from noncentrosymmetric organic molecules, as reported before by several groups [75]. As shown in Figure 8B, in the presence of the cancer cell line, the nature of the plot changes significantly. A polar plot contribution pattern shows clearly four lobes. The asymmetric four-lobe pattern is no longer predominantly dipolar in origin. Therefore, this experimental observation shows that, due to the aggregation in the presence of the cancer cell, one successfully and distinctly observe origin of the very high nonlinearity after the addition of SKBR3 cells is due to the presence of both dipolar and multipolar contributions and could be probed to design nanoparticle based biological sensor for ultra-low-level sensing of pathogens.

CONCLUSION

Nanomaterials became important components in bioanalytical devices since they clearly enhance the performances in terms of sensitivity and detection limits down to single molecule level. Even if long-term toxicity of the nanoparticles is the subject of controversies, the use of plasmonic nanostructures like gold nanoparticles bring many advantages compared with other actual alternatives. Further studies related to the influence of shape and size, capping agents, receptors immobilization onto the metal nanoparticles are still necessary. In this chapter, we reviewed the properties of engineered palsmonic nanostructures their designed size, shape, surface plasmon resonance and surface functionalization for their possibilities to use as probe for biological diagnostics and therapeutics. The aggregation driven surface plasmon resonance shift from blue to near-infrared is important because it allows a better/deeper penetration of the radiation into the body as well as stimulates a drastic change in their spectroscopic signatures and that further could be probed to design a smart sensor. In this brief review, we tried to summarize the most recent development in the field of spectroscopy of nanomaterials in particular Fluorescence spectroscopy, Raman spectroscopy, nonlinear spectroscopy and absorption spectroscopy their application in biology and medicine. Many of techniques highlighted are research-based prototypes not yet ready for integration into commercial device, but they already offer promise when compared to the current gold standards for detection and highlight areas in which nanosensors are able to outperform the traditional detection paradigms. Furthermore, Due to the vast number of different nanomaterials all with its own specific properties, only few examples could be mentioned here by emphasizing the principal advantages of such materials. We provide an overview of some popular nano based approaches illustrate them with current examples showing how they have been applied and discuss their intrinsic benefits and potential labilities.

ACKNOWLEDGMENTS

AKS thanks to Mississippi INBRE, funded by an Institutional Development Award (IDeA) from the National Institute of General Medical Sciences of the National Institutes of Health under grant number P20GM103476. AKS also thanks the Department of Chemistry and

Biochemistry, Jackson State University, for providing access to Analytical core equipment facility funded by NIH under NIH NCRR Grant No. G12RR013459 and NIH/NIMHD Grant No. G12MD007581.

REFERENCES

[1] Thun MJ, DeLancey JO, Center MM, Jemal A, Ward EM. The global burden of cancer: priorities for prevention. *Carcinogenesis.* 2010;31(1):100-10. doi: 10.1093/carcin/bgp263. PubMed PMID: WOS:000273493100011.

[2] Geneau R, Stuckler D, Stachenko S, McKee M, Ebrahim S, Basu S, Chockalingham A, Mwatsama M, Jamal R, Alwan A, Beaglehole R. Chronic Diseases: Chronic Diseases and Development 1 Raising the priority of preventing chronic diseases: a political process. *Lancet.* 2010;376(9753):1689-98. doi: 10.1016/s0140-6736(10)61414-6. PubMed PMID: WOS:000284451000034.

[3] Daraee H, Eatemadi A, Abbasi E, Aval SF, Kouhi M, Akbarzadeh A. Application of gold nanoparticles in biomedical and drug delivery. *Artificial Cells Nanomedicine and Biotechnology.* 2016;44(1):410-22. doi: 10.3109/21691401.2014.955107. PubMed PMID: WOS: 000368022400053.

[4] Wilhelm S, Tavares AJ, Dai Q, Ohta S, Audet J, Dvorak HF, Chan WCW. Analysis of nanoparticle delivery to tumours. *Nature Reviews Materials.* 2016;1(5). doi: 10.1038/natrevmats.2016.14. PubMed PMID: WOS:000377671300001.

[5] Granqvist CG, Buhrman RA, Wyns J, Sievers AJ. Far-infrared absorption in ultrafine al particles. *Physical Review Letters.* 1976;37(10):625-9. doi: 10.1103/PhysRevLett.37.625. PubMed PMID: WOS:A1976CC36900019.

[6] Heinrich U, Fuhst R, Rittinghausen S, Creutzenberg O, Bellmann B, Koch W, Levsen K.Chronic Inhalation Exposure of Wistar Rats and two Different Strains of Mice to Diesel Engine Exhaust, Carbon

Black, and Titanium Dioxide. - *Inhalation Toxicology*. (- 4):- 533. doi: - 10.3109/08958379509015211.

[7] Lam CW, James JT, McCluskey R, Hunter RL. Pulmonary toxicity of single-wall carbon nanotubes in mice 7 and 90 days after intratracheal instillation. *Toxicological Sciences*. 2004;77(1):126-34. doi: 10.1093/toxsci/kfg243. PubMed PMID: WOS:000187988 8000 16.

[8] Erdely A, Hulderman T, Salmen R, Liston A, Zeidler-Erdely PC, Schwegler-Berry D, Castranova V, Koyama S, Kim YA, Endo M, Simeonova PP. Cross-Talk between Lung and Systemic Circulation during Carbon Nanotube Respiratory Exposure. *Potential Biomarkers. Nano Letters*. 2009;9(1):36-43. doi: 10.1021/nl801828z. PubMed PMID: WOS:000262519100007.

[9] Ryman-Rasmussen JP, Tewksbury EW, Moss OR, Cesta MF, Wong BA, Bonner JC. Inhaled Multiwalled Carbon Nanotubes Potentiate Airway Fibrosis in Murine Allergic Asthma. *American Journal of Respiratory Cell and Molecular Biology*. 2009;40(3):349-58. doi: 10.1165/rcmb.2008-0276OC. PubMed PMID: WOS:000263735 700012.

[10] Lindberg HK, Falck GCM, Suhonen S, Vippola M, Vanhala E, Catalan J, Savolainen K, Norppa H. Genotoxicity of nanomaterials: DNA damage and micronuclei induced by carbon nanotubes and graphite nanofibres in human bronchial epithelial cells *In vitro*. *Toxicology Letters*. 2009;186(3):166-73. doi: 10.1016/j.toxlet. 2008.11.019. PubMed PMID: WOS:000265136200005.

[11] Mafune F, Kohno J, Takeda Y, Kondow T, Sawabe H. Formation and size control of sliver nanoparticles by laser ablation in aqueous solution. *Journal of Physical Chemistry B*. 2000;104(39):9111-7. doi: 10.1021/jp001336y. PubMed PMID: WOS:000089712900009.

[12] Kabashin AV, Meunier M. Synthesis of colloidal nanoparticles during femtosecond laser ablation of gold in water. *Journal of Applied Physics*. 2003;94(12):7941-3. doi: 10.1063/1.1626793. PubMed PMID: WOS:000186969900090.

[13] Basak S, Chen DR, Biswas P. Electrospray of ionic precursor solutions to synthesize iron oxide nanoparticles: Modified scaling law. *Chemical Engineering Science.* 2007;62(4):1263-8. doi: 10.1016/j.ces.2006.11.029. PubMed PMID: WOS:000243813900031.

[14] Epifani M, Giannini C, Tapfer L, Vasanelli L. Sol-gel synthesis and characterization of Ag and Au nanoparticles in SiO2, TiO2, and ZrO2 thin films. *Journal of the American Ceramic Society.* 2000;83(10):2385-93. PubMed PMID: WOS:000089834400007.

[15] Ruivo A, Ventura MG, Gomes da Silva MDR, Laia CAT. Synthesis of gold nanoparticles in sol–gel glass porogens containing [bmim] [BF4] ionic liquid. *Journal of Sol-Gel Science and Technology.* 2013;68(2):234-44. doi: 10.1007/s10971-013-3159-6.

[16] Lakshmi BB, Patrissi CJ, Martin CR. Sol−Gel Template Synthesis of Semiconductor Oxide Micro- and Nanostructures. *Chemistry of Materials.* 1997;9(11):2544-50. doi: 10.1021/cm970268y.

[17] Hulteen JC, Martin CR. A general template-based method for the preparation of nanomaterials. *Journal of Materials Chemistry.* 1997;7(7):1075-87. doi: 10.1039/a700027h. PubMed PMID: WOS:A1997XK01100003.

[18] Zhang TJ, Wang W, Zhang DY, Zhang XX, Ma YR, Zhou YL, Qi LM. Biotemplated Synthesis of Gold Nanoparticle-Bacteria Cellulose Nanofiber Nanocomposites and Their Application in Biosensing. *Advanced Functional Materials.* 2010;20(7):1152-60. doi: 10.1002/adfm.200902104. PubMed PMID: WOS:000276775600013.

[19] Singh A, Hede S, Sastry M. Spider silk as an active scaffold in the assembly of gold nanoparticles and application of the gold-silk bioconjugate in vapor sensing. *Small.* 2007;3(3):466-73. doi: 10.1002/smll.200600413. PubMed PMID: WOS:000244909500021.

[20] Song JY, Kim BS. Rapid biological synthesis of silver nanoparticles using plant leaf extracts. *Bioprocess and Biosystems Engineering.* 2009;32(1):79-84. doi: 10.1007/s00449-008-0224-6. PubMed PMID: WOS:000261950900008.

[21] Thakkar KN, Mhatre SS, Parikh RY. Biological synthesis of metallic nanoparticles. *Nanomedicine-Nanotechnology Biology and Medicine.* 2010;6(2):257-62. doi: 10.1016/j.nano.2009.07.002. Pub Med PMID: WOS:000276090600007.

[22] Castro L, Blazquez ML, Gonzalez F, Munoz JA, Ballester A. Extracellular biosynthesis of gold nanoparticles using sugar beet pulp. *Chemical Engineering Journal.* 2010;164(1):92-7. doi: 10. 1016/j.cej.2010.08.034. PubMed PMID: WOS:00028363 3700013.

[23] Reddy AS, Chen CY, Chen CC, Jean JS, Chen HR, Tseng MJ, Fan CW, Wang JC. Biological Synthesis of Gold and Silver Nanoparticles Mediated by the Bacteria Bacillus Subtilis. *Journal of Nanoscience and Nanotechnology.* 2010;10(10):6567-74. doi: 10. 1166/jnn.2010.2519. PubMed PMID: WOS:000280361400043.

[24] Kalishwaralal K, Deepak V, Pandian SRK, Kottaisamy M, BarathManiKanth S, Kartikeyan B, Gurunathan S. Biosynthesis of silver and gold nanoparticles using Brevibacterium casei. *Colloids and Surfaces B-Biointerfaces.* 2010;77(2):257-62. doi: 10.1016/ j.colsurfb.2010.02.007. PubMed PMID: WOS:000276753 600019.

[25] Coker VS, Bennett JA, Telling ND, Henkel T, Charnock JM, van der Laan G, Pattrick RAD, Pearce CI, Cutting RS, Shannon IJ, Wood J, Arenholz E, Lyon IC, Lloyd JR. Microbial Engineering of Nanoheterostructures: Biological Synthesis of a Magnetically Recoverable Palladium Nanocatalyst. *Acs Nano.* 2010;4(5):2577-84. doi: 10.1021/nn9017944. PubMed PMID: WOS:000277976900013.

[26] Singh A, Jain D, Upadhyay MK, Khandelwal N, Verma HN. Green synthesis of silver nanoparticles using argemone mexicana leaf extract and evaluation of their antimicrobial activities. *Digest Journal of Nanomaterials and Biostructures.* 2010;5(2):483-9. PubMed PMID: WOS:000284232000016.

[27] Cui R, Zhang MX, Tian ZQ, Zhang ZL, Pang DW. Intermediate-dominated controllable biomimetic synthesis of gold nanoparticles in a quasi-biological system. *Nanoscale.* 2010;2(10):2120-5. doi: 10.1039/c0nr00193g. PubMed PMID: WOS:000282686200032.

[28] Sweeney RY, Mao CB, Gao XX, Burt JL, Belcher AM, Georgiou G, Iverson BL. Bacterial biosynthesis of cadmium sulfide nanocrystals. *Chemistry & Biology.* 2004;11(11):1553-9. doi: 10.1016/j.chembiol. 2004.08.022. PubMed PMID: WOS:000225613800014.

[29] Hulkoti NI, Taranath TC. Biosynthesis of nanoparticles using microbes-A review. *Colloids and Surfaces B-Biointerfaces.* 2014; 121:474-83. doi: 10.1016/j.colsurfb.2014.05.027. PubMed PMID: WOS:000341335800058.

[30] Senapati D, Singh AK, Ray PC. Real time monitoring of the shape evolution of branched gold nanostructure. *Chemical Physics Letters.* 2010;487(1–3):88-91. doi: http://dx.doi.org/10.1016/j.cplett.2010. 01.015.

[31] Sahoo GP, Bar H, Bhui DK, Sarkar P, Samanta S, Pyne S, Ash S, Misra A. Synthesis and photo physical properties of star shaped gold nanoparticles. *Colloids and Surfaces a-Physicochemical and Engineering Aspects.* 2011;375(1-3):30-4. doi: 10.1016/j.colsurfa. 2010.11.033. PubMed PMID: WOS:000287285500004.

[32] Muller WT, Klein DL, Lee T, Clarke J, McEuen PL, Schultz PG. A strategy for the chemical synthesis of nanostructures. *Science.* 1995;268(5208):272-3. doi: 10.1126/science.268.5208.272. PubMed PMID: WOS:A1995QT 18500032.

[33] Nehl CL, Liao HW, Hafner JH. Optical properties of star-shaped gold nanoparticles. *Nano Letters.* 2006;6(4):683-8. doi: 10.1021/ nl052409y. PubMed PMID: WOS:000236916200017.

[34] Hao F, Nehl CL, Hafner JH, Nordlander P. Plasmon Resonances of a Gold Nanostar. *Nano Letters.* 2007;7(3):729-32. doi: 10.1021/ nl062969c.

[35] Jiang L, Tang YX, Liow CH, Wu JS, Sun YH, Jiang YY, Dong ZL, Li SZ, Dravid VP, Chen XD. Synthesis of Fivefold Stellate Polyhedral Gold Nanoparticles with {110}-Facets via a Seed-Mediated Growth Method. *Small.* 2013;9(5):705-10. doi: 10.1002/smll.201202561. PubMed PMID: WOS:000315769700009.

[36] Huang XQ, Li YJ, Chen Y, Zhou HL, Duan XF, Huang Y. *Plasmonic and Catalytic AuPd Nanowheels for the Efficient*

Conversion of Light into Chemical Energy. Angewandte Chemie-International Edition. 2013;52(23):6063-7. doi: 10.1002/anie.201301096. PubMed PMID: WOS:000319742400031.

[37] Medley CD, Smith JE, Tang Z, Wu Y, Bamrungsap S, Tan W. Gold Nanoparticle-Based Colorimetric Assay for the Direct Detection of Cancerous Cells. *Analytical Chemistry.* 2008;80(4):1067-72. doi: 10.1021/ac702037y.

[38] Darba GKRA, Ray PC. Gold nanoparticle-based miniaturized nanomaterial surface energy transfer probe for rapid and ultrasensitive detection of mercury in soil, water, and fish. *ACS Nano.* 2007;1(3):208-14.

[39] Darbha GKRUS, Singh AK, Ray PC. Selective Detection of Mercury (II) Ion Using Nonlinear Optical Properties of Gold Nanoparticles. *J Am Chem Soc.* 2008;130(25):8038.

[40] Kim HN, Ren WX, Kim JS, Yoon J. Fluorescent and colorimetric sensors for detection of lead, cadmium, and mercury ions. *Chemical Society Reviews.* 2012;41(8):3210-44. doi: 10.1039/C1CS15245A.

[41] Hou JF, Yang YL, Wang C. Molecular Mechanisms of Interface Interactions between Nanomaterials and Proteins. *Acta Physico-Chimica Sinica.* 2017;33(1):63-79. doi: 10.3866/pku.whxb201608233. PubMed PMID: WOS:000392556700005.

[42] Danila DC, Anand A, Schultz N, Heller G, Wan M, Sung CC, Dai C, Khanin R, Fleisher M, Lilja H, Scher HI. Analytic and clinical validation of a prostate cancer-enhanced messenger RNA detection assay in whole blood as a prognostic biomarker for survival. *Eur Urol.* 2014;65(6):1191-7. doi: 10.1016/j.eururo.2013.07.006. PubMed PMID: 23954088; PMCID: PMC4113474.

[43] Chen YX, Huang KJ, Niu KX. Recent advances in signal amplification strategy based on oligonucleotide and nanomaterials for microRNA detection-a review. *Biosensors & Bioelectronics.* 2018;99:612-24. doi: 10.1016/j.bios.2017.08.036. PubMed PMID: WOS:000413284200080.

[44] Beqa L, Singh AK, Khan SA, Senapati D, Arumugam SR, Ray PC. Gold Nanoparticle-Based Simple Colorimetric and Ultrasensitive

Dynamic Light Scattering Assay for the Selective Detection of Pb(II) from Paints, Plastics, and Water Samples. *ACS Applied Materials & Interfaces.* 2011;3(3):668-73. doi: 10.1021/am101118h.

[45] Cao K, Jiang X, Yan S, Zhang L, Wu W. Phenylboronic acid modified silver nanoparticles for colorimetric dynamic analysis of glucose. *Biosens Bioelectron.* 2014;52:188-95. doi: 10.1016/j.bios.2013.08.046. PubMed PMID: 24055932.

[46] Chai FWC, Wang T, Li L, Su Z. Colorimetric detection of Pb2+ using glutathione functionalized gold nanoparticles. *ACS Appl Mater Interfaces.* 2010;2(5):1466-70.

[47] Elghanian R, Storhoff JJ, Mucic RC, Letsinger RL, Mirkin CA. Selective colorimetric detection of polynucleotides based on the distance-dependent optical properties of gold nanoparticles. *Science.* 1997;277(5329):1078-81. doi: 10.1126/science.277.5329.1078. PubMed PMID: WOS:A1997XT03300028.

[48] Luo R, Li YH, Lin XJ, Dong F, Zhang W, Yan L, Cheng W, Ju HX, Ding SJ. A colorimetric assay method for invA gene of Salmonella using DNAzyme probe self-assembled gold nanoparticles as single tag. *Sensors and Actuators B-Chemical.* 2014;198:87-93. doi: 10.1016/j.snb.2014.02.104. PubMed PMID: WOS:000335580100014.

[49] Wang SSAK, Senapati D, Neely A, Yu H, Ray PC. Rapid Colorimetric Identification and Targeted Photothermal Lysis of Salmonella Bacteria by Using Bioconjugated Oval-Shaped Gold Nanoparticles. *Chemistry - A European Journal.* 2010;16(19):5600-6.

[50] Swierczewska M, Lee S, Chen XY. The design and application of fluorophore-gold nanoparticle activatable probes. *Physical Chemistry Chemical Physics.* 2011;13(21):9929-41. doi: 10.1039/c0cp02967j. PubMed PMID: WOS:000290732000003.

[51] Acuna GP, Bucher M, Stein IH, Steinhauer C, Kuzyk A, Holzmeister P, Schreiber R, Moroz A, Stefani FD, Liedl T, Simmel FC, Tinnefeld P. Distance Dependence of Single-Fluorophore Quenching by Gold Nanoparticles Studied on DNA Origami. *Acs Nano.* 2012;6(4):3189-

95. doi: 10.1021/nn2050483. PubMed PMID: WOS:00030309 9300038.

[52] Kang KA, Wang JT, Jasinski JB, Achilefu S. Fluorescence Manipulation by Gold Nanoparticles: From Complete Quenching to Extensive Enhancement. *Journal of Nanobiotechnology*. 2011;9. doi: 10.1186/1477-3155-9-16. PubMed PMID: WOS:000291588100001.

[53] Barzan M, Hajiesmaeilbaigi F. Fluorescence quenching of Rhodamine 6G with different concentrations by laser ablated gold nanoparticles. *Optical and Quantum Electronics*. 2015;47(11):3467-76. doi: 10.1007/s11082-015-0222-2. PubMed PMID: WOS: 000362578800004.

[54] Tira DS, Focsan M, Ulinici S, Maniu D, Astilean S. Rhodamine B-Coated Gold Nanoparticles as Effective "Turn-on" Fluorescent Sensors for Detection of Zinc II Ions in Water. *Spectroscopy Letters*. 2014;47(2):153-9. doi: 10.1080/00387010.2013.782557. PubMed PMID: WOS:000328599800009.

[55] Chang HY, Hsiung TM, Huang YF, Huang CC. Using Rhodamine 6G-Modified Gold Nanoparticles To Detect Organic Mercury Species in Highly Saline Solutions. *Environmental Science & Technology*. 2011;45(4):1534-9. doi: 10.1021/es103369d. PubMed PMID: WOS:000287122400054.

[56] Dulkeith E, Morteani AC, Niedereichholz T, Klar TA, Feldmann J, Levi SA, van Veggel F, Reinhoudt DN, Moller M, Gittins DI. Fluorescence quenching of dye molecules near gold nanoparticles: Radiative and nonradiative effects. *Physical Review Letters*. 2002;89(20). doi: 10.1103/PhysRevLett.89.203002. PubMed PMID: WOS:000178938500015.

[57] Fan Z, Dai X, Lu Y, Yu E, Brahmbatt N, Carter N, Tchouwou C, Singh AK, Jones Y, Yu H, Ray PC. Enhancing targeted tumor treatment by near IR light-activatable photodynamic-photothermal synergistic therapy. *Mol Pharm*. 2014;11(4):1109-16. doi: 10.1021/mp4002816. PubMed PMID: 24568338; PMCID: PMC3983349.

[58] Huang XESIH, Qian W, El-Sayed MA. Cancer Cell Imaging and Photothermal Therapy in the Near-Infrared Region by Using Gold Nanorods. *J Am Chem Soc.* 2006;128(6):2115-20.

[59] Cao YC, Jin RC, Nam JM, Thaxton CS, Mirkin CA. Raman dye-labeled nanoparticle probes for proteins. *Journal of the American Chemical Society.* 2003;125(48):14676-7. doi: 10.1021/ja0366235. PubMed PMID: WOS:000186834500013.

[60] Dieringer JA, Wustholz KL, Masiello DJ, Camden JP, Kleinman SL, Schatz GC, Van Duyne RP. Surface-Enhanced Raman Excitation Spectroscopy of a Single Rhodamine 6G Molecule. *Journal of the American Chemical Society.* 2009;131(2):849-54. doi:10.1021/ja8080154. PubMed PMID: WOS:000262521800079.

[61] Fan Z, Senapati D, Khan SA, Singh AK, Hamme A, Yust B, Sardar D, Ray PC. Popcorn-shaped magnetic core-plasmonic shell multifunctional nanoparticles for the targeted magnetic separation and enrichment, label-free SERS imaging, and photothermal destruction of multidrug-resistant bacteria. *Chemistry.* 2013; 19(8):2839-47. doi: 10.1002/chem.201202948. PubMed PMID: 23296491.

[62] Senapati D, Dasary SSR, Singh AK, Senapati T, Yu H, Ray PC. A Label-Free Gold-Nanoparticle-Based SERS Assay for Direct Cyanide Detection at the Parts-per-Trillion Level. *Chemistry – A European Journal.* 2011;17(30):8445-51. doi: 10.1002/chem. 201100617.

[63] Singh AK, Khan SA, Fan Z, Demeritte T, Senapati D, Kanchanapally R, Ray PC. Development of a Long-Range Surface-Enhanced Raman Spectroscopy Ruler. *Journal of the American Chemical Society.* 2012;134(20):8662-9. doi: 10.1021/ja301921k.

[64] Khan SA, Singh AK, Fan Z, Senapati D, Ray PC. Designing distance dependent SERS assay for monitoring photothermal antibacterial activity response. *Chemical Communications.* 2012;48(90):11091-3. doi: 10.1039/c2cc36147g. PubMed PMID: WOS:000309918000009.

[65] Yang MH, Kostov Y, Bruck HA, Rasooly A. Gold nanoparticle-based enhanced chemiluminescence immunosensor for detection of

Staphylococcal Enterotoxin B (SEB) in food. *International Journal of Food Microbiology.* 2009;133(3):265-71. doi: 10.1016/j.ijfoodmicro.2009.05.029. PubMed PMID: WOS:000269115200 008.

[66] Temur E, Zengin A, Boyaci IH, Dudak FC, Torul H, Tamer U. Attomole Sensitivity of Staphylococcal Enterotoxin B Detection Using an Aptamer-Modified Surface-Enhanced Raman Scattering Probe. *Analytical Chemistry.* 2012;84(24):10600-6. doi: 10.1021/ac301924f. PubMed PMID: WOS:000312429800016.

[67] Lee HM, Jin SM, Kim HM, Suh YD. Single-molecule surface-enhanced Raman spectroscopy: a perspective on the current status. *Physical Chemistry Chemical Physics.* 2013;15(15):5276-87. doi: 10.1039/C3CP44463E.

[68] Damborska D, Bertok T, Dosekova E, Holazova A, Lorencova L, Kasak P, Tkac J. Nanomaterial-based biosensors for detection of prostate specific antigen. *Microchimica Acta.* 2017;184(9):3049-67. doi: 10.1007/s00604-017-2410-1. PubMed PMID: WOS:000407 371700002.

[69] Dasary SSR, Singh AK, Senapati D, Yu H, Ray PC. Gold Nanoparticle Based Label-Free SERS Probe for Ultrasensitive and Selective Detection of Trinitrotoluene. *Journal of the American Chemical Society.* 2009;131(38):13806-12. doi: 10.1021/ja905134d.

[70] Nie SM, Emery SR. Probing single molecules and single nanoparticles by surface-enhanced Raman scattering. *Science.* 1997; 275(5303):1102-6. doi: 10.1126/science.275.5303.1102. PubMed PMID: WOS:A1997WJ50300037.

[71] Lu Wasa, Senapati D, Singh AK, Arbneshi, T, Khan Sayh, Ray PC. Multifunctional Oval-Shaped Gold-Nanoparticle-Based Selective Detection of Breast Cancer Cells Using Simple Colorimetric and Highly Sensitive Two-Photon Scattering Assay. *ACS Nano.* 2010;4(3):1739-49.

[72] Kim JH, Park JS, Kim MG, editors. Strong Hyper-Rayleigh scattering from silver nanoparticle aggregates to be used for the optical bio-sensing assay. *Conference on Nano-Bio Sensing,*

Imaging, and Spectroscopy (NBSIS); 2013 Feb 20-23; Korean Sensors Soc, Jeju, South Korea 2013.

[73] Zhang CX, Zhang Y, Wang X, Tang ZM, Lu ZH. Hyper-Rayleigh scattering of protein-modified gold nanoparticles. *Analytical Biochemistry.* 2003;320(1):136-40. doi: 10.1016/s0003-2697(03) 00353-1. PubMed PMID: WOS:000184794200014.

[74] Shen H, Iop, editors. Plasmonic nanoparticles: Towards the fabrication of biosensors. *Global Conference on Polymer and Composite Materials* (PCM); 2015 May 16-19; Beijing, Peoples R China 2015.

[75] Johnson RC, Li JT, Hupp JT, Schatz GC. Hyper-Rayleigh scattering studies of silver, copper, and platinum nanoparticle suspensions. *Chemical Physics Letters.* 2002;356(5-6):534-40. doi: 10.1016/ s0009-2614(02)00407-4. PubMed PMID: WOS:000175379000017.

[76] Singh AK, Senapati D, Wang S, Griffin J, Neely A, Candice P, Naylor KM, Varisli B, Kalluri JR, Ray PC. Gold Nanorod Based Selective Identification of Escherichia coli Bacteria Using Two-Photon Rayleigh Scattering Spectroscopy. *ACS Nano.* 2009;3(7):1906-12. doi: 10.1021/nn9005494. PubMed PMID: 19572619; PMCID: PMC2888863.

In: Nanotechnology
Editor: Paresh Chandra Ray
ISBN: 978-1-53613-889-4

Chapter 4

NANOPARTICLES FOR PHOTOTHERMAL THERAPY IN CANCER

Aruna Vangara, Avijit Pramanik, Salma Begum, Kaelin Gates, Ye Gao and Paresh Chandra Ray*
Department of Chemistry and Biochemistry,
Jackson State University, Jackson, Mississippi US

ABSTRACT

Cancer treatment has been developed over years with wide contributions of research work. Photothermal therapy has its preference among the cancer research communities for its specific, accurate, precise treatment of cancer cells while less damage to healthy cells. This treatment also has additional advantage of selectively treating surface and deep tumors. For deep tumors the nanoparticles should be efficiently absorbing near infrared light mostly 808 nm to generate heat. The following chapter discusses about the widely known nanoparticle shape, size and materials and its contributions in photothermal therapy in cancer. It is widely known that nanoparticles has finely tuned surface plasmon

* Corresponding Author Email: paresh.c.ray@jsums.edu. Fax: +16019793674.

resonance and thus finely tune to absorb near infrared light. Initially we discussed about different shapes and sizes of nanoparticles, nanoshell, nanorods and nanowires such as copper nanowires, tellurium platinum nanowires and silica nanowires followed by different materials such as inorganic nanoparticles such as bismuth sulfide nanoparticles, copper sulfide nanoparticles, silica nanoparticles, carbon nanotubes and polymeric nanoparticles. Each of the research work discussed in this chapter has several different sizes, near infrared light, near infrared light energy and great potential for cancer treatments. The research work in this chapter has additional advantages such as multifunctional nanoparticles such as photodynamic therapy with reactive oxygen species generation, fluorescence imaging, photoacoustic imaging and optical coherence tomography to give synergistic cancer treatment.

Keywords: nanoparticles, inorganic nanoparticles, polymeric nanoparticles, near infrared light, multifunctional nanoparticles

Introduction

Cancer is the second leading cause of health concern in the United States [1-10]. According to the American Cancer Society, about 1,688,780 new cancer cases are expected to be diagnosed in 2017 and 600,920 cancer deaths. Both these numbers have risen compared to 2016 [7-8]. According to the National Cancer Institute, in the next two decades, it is estimated that the number of cases and number of cancer deaths are estimated to increase by 50% and 60%, respectively worldwide [7-8]. According to the National Cancer Institute, the overall cancer deaths rates decreased for men, women and children and it was also observed that there is significant improvement in the survival rate [7-8]. The main reason for this improvement is due to many significant advances over the years in the targeted treatment and diagnosis but even today cancer remains a serious health concern [7-8].

The widely used treatments in cancer are surgery, radiation therapy, chemotherapy, immunotherapy and stem cell therapy [11-30]. These treatments are given individually or in combination. Although widely used, these treatments are expensive and are associated with risks and serious side effects like infections from invasive treatments, damage to healthy cells, ineffective early stage detection and diagnosis of circulating tumor cells, metastasis and ability to spread [1-20]. Besides these side effects, chemotherapy can lead to drug resistance and radiation therapy can destroy normal cells [1-10]. Also, these side effects may or may not improve and may recur after months or years after treatment. Besides these side effects, cancer spread in the body through circulating tumor cells (CTCs) leading to tumor metastasis. Hence there is a huge need for a highly sensitive and precise treatment which is less invasive and gives accurate diagnosis and effective therapeutic results with minimal to no side effects [31-50].

A promising therapeutic method that is gaining much attention recently among the cancer research communities worldwide is photothermal therapy (PTT), a treatment highly selective in killing cancer cells while less damaging to healthy cells. PTT is a non invasive treatment, has minimal to low toxicity and efficient bioavailability and can be a potential alternative to the invasive treatments [30-70]. Technically, PTT can be also referred as hyperthermia. In Figure 1 shows a schematic representation of Photothermal therapy using photothermal agent Copper Nanowires (CuNWs) stabilized with polyethylene glycol (PEG)*in vivo* and irradiated with 808nm NIR light [71-94].

In this method, cancer cells are exposed to a photothermal agent using nanoparticles, and then irradiated with laser light of appropriate wavelength (usually near infrared light of wavelength 650nm to 900nm) on the area exposed to photothermal agent [31-60]. Near infrared light (NIR) is most commonly used in photothermal therapy because it can reach deep tumors, about 10cms, depending on the type of tissue and also less interaction with surrounding environment such as water and hemoglobin [60-94]. Tumors usually have porous vasculature/blood vessels and this porous nature becomes more prominent with the heat generated from the photothermal agent thus enhances the bioavailability and accumulation of photothermal agents inside the cells for efficient therapeutic efficacy [20-40]. The photothermal agent absorbs visible to near infrared light. This absorbed light is converted into heat through nonradiative pathways [31-60]. This highly localized heat can cause irreversible damage to cancer cells that usually involves changes to cell adhesion and morphology or inhibition of cell proliferation [20-40]. Thus, photothermal agents efficiently destroy cancer cells. The nanoparticles with high optical absorption capabilities and provide combined photodynamic therapy (PDT)/photothermal therapy (PTT)/fluorescence imaging/chemotherapy has recently gained much attention [31-50]. Such nanoparticles are reported as highly promising therapeutic agents which prevents cancer metastasis [31-50]. Here we discuss few of the most extensively researched and fascinating examples of nanophotothermal agents based on different types of nanomaterials, their shape, size and their perspective in photothermal therapy in cancer and multifunctionality.

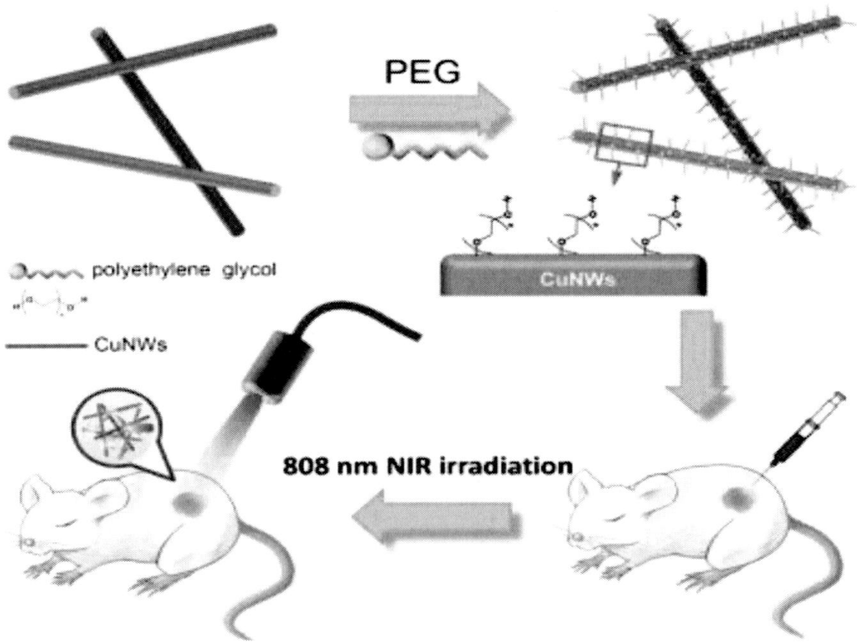

Figure 1. Schematic representation of Photothermal therapy using copper nanowires. Adapted with permission from *ACS Appl. Mater. Interfaces*, 2016, 8, 12082-12090 [2].

GOLD NANOPARTICLES AS NANOPHOTOTHERMAL AGENTS

Nanospheres

Due to surface plasmon resonance, gold nanoparticles can absorb and scatter the light and can efficiently generate heat for photothermal therapy [1-6, 10-20]. When gold nanoparticles absorb light of appropriate wavelength, the electrons excite. These excited electrons return to ground state by releasing the energy as heat. This heat increases temperature of the surrounding area, which is high enough to cause cancer cell shrinkage/ necrosis and thus treats cancer cells [1-6, 10-20]. In detail, the free electrons or the conducting electrons on the surface of the gold nanoparticles absorb the light of certain wavelength and oscillate with a specific frequency. Then the energy of the free electrons increases and electrons gets excited [1-6, 10-20]. These excited electrons come to ground state by releasing the absorbed energy as heat. This electronic property known as surface plasmon resonance (SPR), Figure 2.

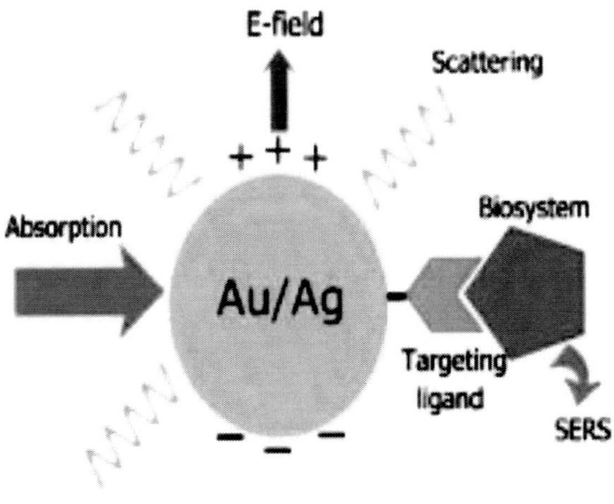

Figure 2. The schematic representation of Gold nanoparticles or Silver nanoparticles attached to a ligand absorb and scatter light. Adapted with permission from *Acc. Chem. Res.*, 2008, 41, 1578-1586 [5].

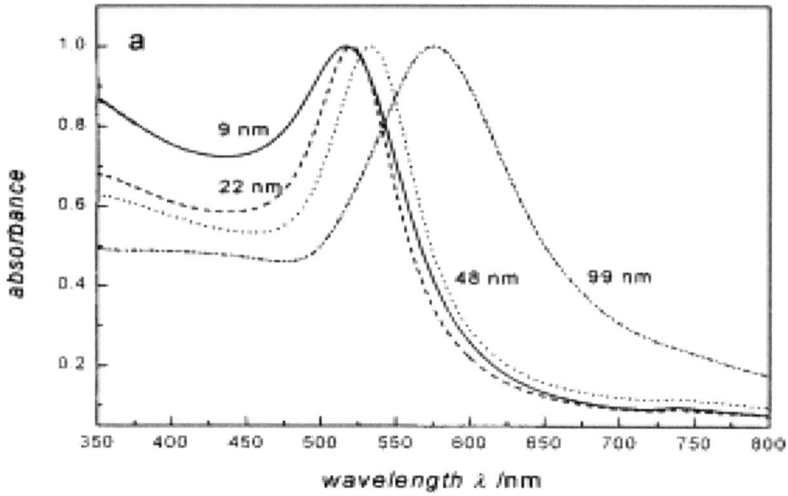

Figure 3.1. Increasing gold nanoparticle size and the red shift in plasmon absorbance.

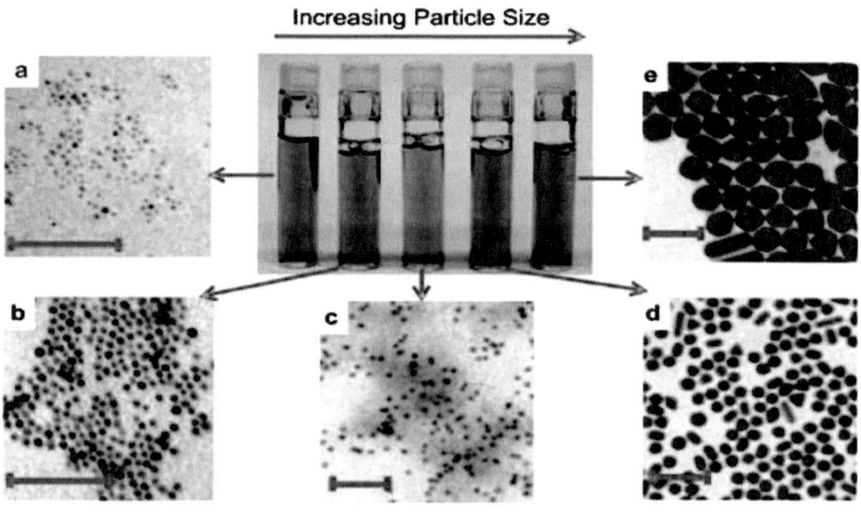

Figure 3.2. TEM and colors of Gold nanoparticles scattering light changes with increasing size. Adapted with permission from *Acc. Chem. Res.* 2008, 41, 1721-1730 [5].

Nanoparticles have absorption and scattering of light depending on size, shape, surrounding medium that gives the unique advantage in cancer diagnosis, imaging and photothermal therapy [1-6, 20-30]. By tuning ratio of absorption to scattering in gold nanoparticles the therapeutic function can be tuned. Larger particle has larger surface area provides high scattering efficiency to be used as contrast agent in scattering based imaging techniques such as tissue imaging using reflectance confocal microscopy and optical coherence tomography and small nanoparticles has high absorption efficiency that generate heat for photothermal therapy applications [1-6, 20-30]. For example, gold nanoparticles with larger sizes more scattering compared to absorption and more red shift in the wavelength compared to small size gold nanoparticles shows intense light absorption at approx. 522nm as shown in spectra Figure 3 and Figure 4.

To understand the efficiency of gold nanoparticles, it was calculated that a single gold nanoparticle has scattering intensity same to that of emission intensity from 10^4 fluorescein molecules [1-6, 20-30]. For example, a 60nm gold nanoparticle has intense light scattering about 10^5 times stronger than an organic dye emission such as fluorescein. Due to surface plasmon resonance and higher absorbance efficiency in nanoparticles than the conventional dyes, nanoparticles have a promising role in hyperthermia [1-6, 20-30]. Near infrared nanoheaters have greater advantages of non photobleach, high photostability, better sensitivity and better biostability compared to conventional near infrared organic dyes [1-6, 20-30]. Gold nanoparticles also contribute in conventional cancer treatments. Smilowitz et al. improved conventional radiation therapy by reducing the dose required using 15nm gold nanoparticles through x-ray heating process [1-6, 30-40]. This research allows gold nanoparticles to reduce the dose related side effects of radiotherapy. Thus the therapeutic efficiency of conventional radiation therapy is increased with less side effects [1-6, 20-30].

Nanoparticles with sufficient circulation time to accumulate at tumor site and high specific cancer cell interaction are important for their application in *in vivo* cancer diagnosis and treatment [1-6, 20-30]. The nanoparticles reaches tumor either passive or active. Nanoparticles accumulate at tumor passively due to high porous blood vessels and less lymphatic drainage of tumor [1-6, 30-40]. Nanoparticles accumulate actively at tumor site by surface functionalization with several different molecules like antibodies, peptides for tumor specific interaction [1-6, 20-

30]. Surfactants are commonly used to improve steric and electrostatic stabilization. In general, these surface modifications improve the solubility, stability by preventing aggregation, bioavailability and also enhance therapeutic effectiveness in drug delivery/imaging and photothermal therapy [1-6, 40-50]. Chen et al. developed gold nanoparticles coated with poly(perylenediimide) (PPDI) and poly(ethylene glycol) (PEG). This gold nanoparticle coated with PEG and PPDI nanovesicle demonstrated enhanced absorption, photoacoustic intensity and photothermal therapy by controlling interparticle distance [46]. In order to treat radiation resistant tumor cells, Xing et al. developed glucose attached gold nanoparticles that improves cancer cells sensitivity to radiation [47].

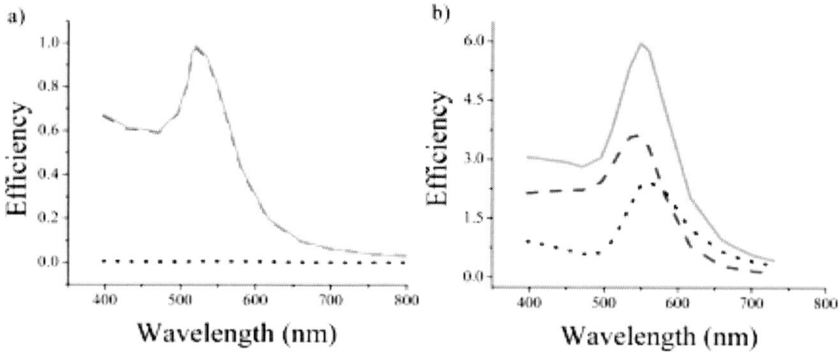

Figure 4. The extinction spectra (green solid line), absorption spectra (red line) and scattering spectra (black line) for A) 20nm gold nanoparticle and B) 80nm gold nanoparticle. Adapted with permission from *J. Phys. Chem. B*, 2006, 110, 7238-7248 [45].

Generally, gold nanoparticles has high visible light absorption and comparatively less near infrared light absorption. Near infrared light absorption is highly preferable for applications *in vivo*. Thus varying the shape and structure of nanoparticles to nanoshells, nanorods and by varying materials to inorganic or polymeric nanoparticles unique properties of nanoparticles are explored to understand the nanosystem efficiency in photothermal therapy [1-6, 20-30]. Gold nanospheres usually have absorbance at UV or visible light which has limitations in photothermal therapy for deep tumors but can be applied for surface cancers like skin [1-

6, 20-30]. El Sayed et al. showed that gold nanoparticles ~40nm size with surface plasmon absorbance at visible light 530nm result in hyperthermia and irreversible cancer cell damage [3-5]. They also demonstrated that the malignant cancer cells at less than half the energy required to damage benign HaCaT cells. the importance of biomolecules attached to nanoparticles in both enhancing the specificity, accumulation of more concentration of nanoparticles at the cancer cells and thus generates more heat with low energy light for photothermal therapy of cancer. These biomolecules causes specific attachment to a cell type thus increasing the concentration of nanoparticles at the cell surface or within the cell. The two malignant cells which are incubated with anti EGFR attached gold nanoparticles then exposed to visible argon ion light 514nm shows that HOC cells requires less energy compared to HSC cells [3-5, 11]. This difference is due to higher accumulation of anti EGFR attached gold nanoparticles in HOC cells than HSC cells. Compared to malignant cells, the benign HaCaT cells requires more than twice the energy to undergo irreversible cell damage to benign HaCaT cells than that was required in HOC and HSC cells [3-5, 11]. The benign and malignant cells without nanoparticles remain viable even when irradiated with high power laser $76W/cm^2$ for 4 mins can be used to absorb visible argon ion light at 514nm [11].

Besides enzyme responsive mechanism to have more accurate treatment of cancer via photothermal therapy, FRET mechanism is also explored with nanoshells [1-6, 40-50]. This FRET based theranostic nanoparticle allows non invasive real time simultaneous imaging of tumor and precise cancer treatment. In FRET, there is an energy transfer between donor and acceptor, thus obtains high contrast images compared to conventional imaging techniques [1-6, 40-50]. The main feature for the efficient FRET is that the distance between the two components, i.e., donor and acceptor is 1nm-10nm and the fluorescence emission of donor overlaps with absorbance of acceptor. An efficient FRET nanoparticles was designed with widely accepted dyes, Methylene blue (MB) acts as donor and Indocyanine green (ICG) acts as acceptor where the distance between them is 1nm-2nm and has good overlap of absorbance/fluorescence [1-6,

40-50]. This nanoparticle *in vitro* and *in vivo* allow dual imaging, i.e., fluorescence and photoacoustic imaging. Also these nanoparticles in C6 tumor ~60 mm^3 size when activated with near infrared light 808nm, 1.0 W/cm^2, 5 min. shows 94.8% of therapeutic efficacy against C6 glioma cells with 100% survival rate [48].

Several groups developed hollow gold nanosphere consists of hollow gold core coated with thin layer of polyethylenimine (PEI) and finally attached to hydrophobic ICG [1-6, 40-50]. The amino group of PEI is covalently conjugated to sulfonic acid group of ICG. This strong interactions between PEI and ICG gives excellent stability at different pH, greater intracellular stability, enhanced fluorescence quantum yield, approx. < 10% ICG leakage during *in vivo* circulation, enhanced fluorescence, synergistic effect both PDT/PTT and enhanced plasmon absorption in NIR region thus enhanced PTT compared to hollow gold nanospheres (HAuNS) or free ICG itself [1-6, 80-90]. A slightly lower PDT efficacy was noticed due to ROS productivity was closely related to the fluorescence of ICG resulting in a decreased ROS productivity under NIR laser irradiation. Compared to free ICG, HAuNS, ICG-HAuNS, the ICG conjugated PEI HAuNS provides excellent Anti tumor efficacy was on CT-26 tumors and anti metastasis efficacy on B16 tumor with synergistic PTT/PDT when irradiated by NIR light by killing tumor cells and blood vessels to tumor [90].

Nanoshells

Earlier we have discussed single component solid nanoparticles but in the case of nanoshells, there are two materials to give nanoshell nanoparticles [50-60]. When these two materials are placed in concentric structures, these nanostructures have greater preference over the solid nanoparticles because of advanced functionalities like stronger red shift and also these structures have pin holes that also enhance the Surface Enhanced Raman Spectroscopy (SERS) compared to the solid component nanoparticle [50-60]. Thus these nanostructures are more preferred than the solid component nanoparticles [50, 51]. In this, core is usually made up of nonmetallic material and shell is made up of a metallic material and vice versa. These cores surround by thin coatings of shell giving a concentric morphology. Similar to single component nanoparticles, we can tune the excitation wavelength that causes electron oscillation in nanoshells by changing the material or by changing the size, i.e., core to shell ratio [50-60]. Nanoshell consists of dielectric silica inner core and a metallic outer shell. When nanoshell absorbance overlaps with the NIR laser light emission wavelength, converts the absorbed light into heat. The first nanoshell was designed with silica core within the gold shell by Halas et al. [52]. The unique properties of nanoshell are tunable surface plasmon resonance and by varying the dimensions of the core to shell in nanoshells can result in change in the colors nanoshell solution [50-60]. Size, generally less than 100nm is considered to be ideal. Since large particles have poor clearance and can be toxic while small particles less than 5nm has poor bioavailability due to renal clearance [54]. With a biocompatible surface modification, the nanoparticles have good circulation, biodistribution, good bioavailability and thus allowing nanoshells as promising efficient therapeutic agents [50-60]. Adding functionalities to the nanoshell which has plasmonic shell and non plasmonic core and vice versa with bioagents to enhance the specificity to tumor cells can enhance the size. Hence controlling the size is not easy [50-60]. According to Mie theory, the plasmon resonance shifts red when the thickness is decreased in nanoshells and the size is increased [50]. It was demonstrated that increasing the core to shell ratio from 3nm to 12nm can red shift to about 300nm, i.e., from visible to infrared of the spectrum [53].

Among different ligands attached to nanoshells, fluorophore attached nanoshells are widely known for enhanced fluorescence efficiency [50-70]. As it is widely known that fluorophores have photobleach, this can be avoided by attaching nanoshells, a concept known as metal enhanced fluorescence [52]. The size of the metal nanostructures determines the fluorescence of the molecular fluorophore to be enhanced or quench. Small size nanoagents can quench while large size nanoagents can enhance the fluorescence. Compared to conventional NIR dyes like ICG, the nanoshells has million fold highly effective NIR absorbers [61]. Thus Nanoshells are found to be potential in photothermal therapy. Nanoshell combined therapy such as image guided cancer treatment can be obtained by using the fluorophore, as contrast agent and in drug delivery [50-60]. The conventional nanoshells have disadvantage of nonspecific biodistribution which can be resolved by using different ligands and anticancer drugs that enhances the specificity to tumor cells [50-60]. Adding to specificity these ligands also contributes to theranostic functionality such as simultaneous diagnosis and treatment [50-60]. Yongdong et al. combined the photoacoustic effect of gold nanoshells (AuNSs) functionalized by using different core molecules to give a new class of multifunctional hybrid AuNSs that enables biosensing, multimodal imaging and controlled release. They also worked on enzyme responsive silver/gold bimetallic nanoshells as biosensor [54].

Compared to single modal application of nanoparticles, multi-functionality is always preferred that are highly stable in aqueous solution and intracellular [50-60]. A polyethylene glycol (PEG) coated nanoshell has enhanced stability to multifunctional nanoshells for *in vitro* and *in vivo* cancer treatment. For example, silica/Au nanoshell with 120nm core and 10nm surface modified with antiHER2 and PEG was synthesized to specifically attach to SKBR3 breast cancer cells [50-60]. These modified nanoshells allow combined imaging and photothermal therapy. The imaging of these nanoshells was carried out after 1hr incubation with SKBR3 cells. It was found that there is a significant increase in contrast images observed in scattering based microscope [50-60]. The nanoshell PEGylation enhances the *in vivo* and *in vitro* stability of nanoparticle without affecting the plasmon resonance peak [55]. In *in vitro* study the antiHER2 attached nanoshells were added to SKBR3 cells. Then these cells were imaged followed by NIR irradiation at 820 nm at a low power 0.008 W/m^2 for 7 min. These nanoshells with NIR light treated cells were assessed for nanoshell specificity to the SKBR3 cell surfaces and allow hyperthermia cell damage [55]. It was observed that the nanoshells were specifically attached to the SKBR3 cells and provides enhanced contrast to the scattering based imaging technology. The fluorescence images obtained shows significant cell damage in antiHER2 attached nanoshells and NIR laser treated cells whereas the cells treated with antiHER2 attached nanoshells without NIR light has no cytotoxicity. Thus these nanoshells designed to provide both diagnosis and therapeutic [55]. Because of ease of surface functionalization with ligands the sensitivity to NIR laser or to tumor acidic pH can be manipulated to establish nanoshells application in site specific drug/ molecule delivery [50-80]. For example, Hollow nanostructures have gained considerable interest from researchers due to unique electronic properties, large surface area, large drug loading capacity, low density, high absorption coefficient ranging from visible to NIR region and less harmful to healthy tissues compared to nanoshells [91]. Instead of solid silica core, hollow silica core can enhance the amount of hydrophilic or hydrophobic drug/molecule content and thus provides additive chemophototherapy to cancer [56]. Hollow silica (SiO_2)gold nanoshells with hydrophilic drug exhibits control release of drug and broad NIR absorption results in simultaneous chemotherapeutic and photo-therapeutic efficacy [56].

Most research considers silica dielectric as core but it was found that silica as shell can enhance the stability in hollow porous silica shell gold/ silver nanoshells compared to PEG coated hollow nanoshells at low acidic pH [10-20]. Gold core nanoshells are comparatively non toxic to silica core nanoshells [11]. To determine the stability of hollow silica gold/silver nanoshells, a study over wide range of pH 2 to 10 was conducted. These pH ranges was considered to represent the various biological pH conditions. Porous silica coated hollow gold/silver nanoshells size approx. [57]. 110nm was synthesized owing their enhanced stability to silica coating. Initially, silver nanoparticles are synthesized as core then coated with silica to enhance stability to give porous silica coated silver nanoparticles and then finally added gold solution to give porous silica coated hollow gold/silver nanoshells [57]. It was observed that during synthesis when porous silica coated silver nanoparticles is converted to porous silica gold/silver nanoshells the SPR peak was red shifted to NIR region. At pH 7.0 has UV Vis absorbance peak is at approx [57]. 800nm and when pH > 7.0 the peak is blue shifted to ~770nm and when pH = 4.0

acidic shows slight red shifted and when more acidic more red shift was observed. In their research work they also showed that these hollow nanoshells SPR absorbance tuned with change in pH. It was also observed that the nanoshells remain intact for a wide range of pH conditions [57]. Additionally photothermal study was conducted by irradiation of these nanoparticles at 808nm for 10 mins and observed a significant temperature change [57]. A bimetallic core shell type nanoparticles were developed due to broad absorption band ranging from visible to NIR region [57]. Typically these bimetallic core shell nanoparticles consist of silver core surrounded by gold or gold/silver alloy shell. A 40nm to 100 nm ultra small spherical hollow nanoshells made up of hollow silver core and gold shell were prepared [57]. It was demonstrated that by controlling the size these nanoshells, a large red shift with extinction maxima at 950nm to 1000nm can be obtained which is primarily because of gold shell [58]. Another group developed drug loaded Au/Ag hollow nanoshells with acidic pH and NIR light responsive drug release [59]. The nanoshell consists of silver nanoparticles to form inner hollow shell and gold as outer thin layer. They showed that these nanoshells are both acidic and NIR light promotes the drug release [59]. They also demonstrated the passive and active tumor accumulation of nanoshells [59]. To demonstrate the efficiency of passive and active tumor accumulation, the drug loaded/EGFR attached Au/Ag hollow nanoshells were treated with lung cancer cells for 2hrs at room temperature [59]. Then for passive targeting the nanoshell treated cancer cells were incubated for 4hrs prior irradiation for endocytosis tumor accumulation then irradiated at 800nm continued to incubate for 12hrs whereas for active targeting the nanoshell treated cancer cells were irradiated and then incubated for 16hrs [59]. The active approach allows more nanoshells to enter the tumor environment compared to the passive approach for lung cancer treatments. Thus these nanoshells NIR irradiated provide simultaneous synergistic combined chemotherapy and photo-thermal therapy [59].

Metal nanoshells are often debatable for *in vivo* metal provides systemic toxicity due to combined aggregation and accumulation while metal nanoshells has significant SPR effect and large NIR absorption from

800nm to 1300nm where tissues absorb less and transmit more to induce deep tissue anti tumor effect [45-60].

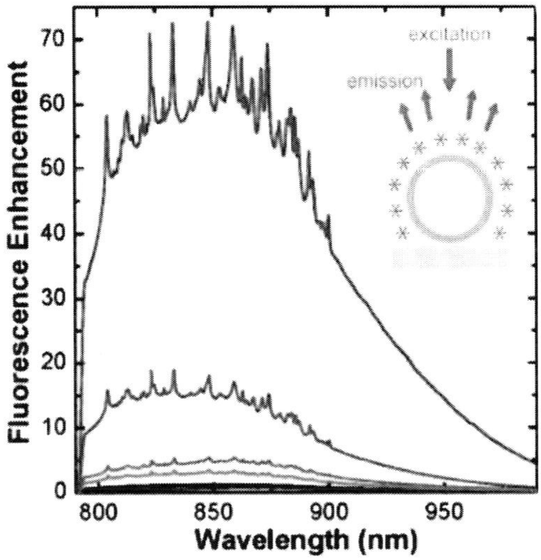

Figure 5. The ICG emission enhanced with nanoshell core size 112nm and shell 123nm compared to nanoshells with core size 64nm/112nm and shell 88nm, 81nm, 130nm respectively. Adapted with permission from *Nano Lett.*, 2007, 7, 496-501 [62].

Nanoshells dimensions controlled and conjugated with antibody can obtain desired results expanding their therapeutic applications [45-60]. After antibody conjugation, the surface of gold nanoshells further coated with PEG-thiol to prevent non specific interactions with molecules other than antibody [45-60]. The nanoshell can be finely tuned and measure for scattering and absorption is well known. Nanoshells have larger absorption surface area compared to nanospheres/ conventional organic dyes and less photobleachable, easily conjugated to ligands and thus enhanced bioavailability compared to organic dyes [45]. Large surface area means more scattering which enhance the nanoshells applications in scattering based imaging techniques [45-60]. While this scattering property allows detection/diagnosis by obtaining real time/high resolution scattering images using optical coherence tomography and the absorption property allows treating to inhibit cancer cells/metastasis [45-60]. Based on core to shell thickness, the absorbing nanoshells and scattering nanoshells were developed. It was demonstrate that the scattering nanoshells as contrast agent for *In vivo* imaging by using optical coherence tomography and the absorbing nanoshells for photothermal therapy [61]. For this purpose the group developed large size 100nm core and 20nm shell thickness for optical coherence tomography imaging and small size 60nm core and 10nm shell thickness for photothermal therapy [61]. The scattering efficiency was also demonstrated, the nanoshell solution in water is prepared and obtained optical coherence tomography images at 830nm [61]. To demonstrate *in vitro* photothermal application two groups of breast cancer cells were treated with PEG-Gold nanoshells of size ~110 nm diameter core and a 10-nm-thick shell with absorbance at 820 nm. In this two groups of breast cancer cells, one group treated with nanoshells + laser and another control group [60-61]. The group treated with nanoshells and laser at 820nm and high laser dose ~35W/cm^2 for 7 mins shows significant cell shrinkage and necrosis whereas control group as without nanoshells and only NIR laser shows no significant cell damage. For *in vivo* low doses of NIR light approx. 4W/cm^2 in presence of metal nanoshells gives maximum temp. 37.4 ± 6.6°C rise in tumor with 4-6 mins [60-61]. Followed by photothermal therapy, fluorescence imaging was performed and observed a significant cell shrinkage and irreversible cell damage [60]. Among fluorophores such as ICG, IR800 with low quantum yield and highly photobleachable and thus unstable whose poor emitting properties can be altered when conjugated with a metallic nanoparticle due to plasmon resonance of nanoparticles and its scattering efficiency, i.e., large scattering efficiency of large size nanoshells and plasmon resonance of nanoshells overlaps with $\lambda_{emission}$ of fluorophore [60-61]. Another group showed that nanoshells with 112nm core size and 123nm shell has $\lambda_{emission}$ overlaps with ICG molecule, ICG emission enhanced to > 50 fold [62] compared to core size 64nm/112nm and shell 88nm, 81nm, 130nm respectively as shown in Figure 5 [62].

A direct adsorption of fluorophore to nanoparticle can quench the fluorescence but when applied a 5-11nm layer such as HSA (human serum albumin) can prevent quenching and enhances the fluorescence emission [60-62]. Nanoshells core size 63nm and shell 78nm and nanorods with width 11nm and length 46nm were coated with HSA (Human Serum Albumin) approx. 8nm it was observed that the Nanoshell/IR800 with high fluorescence intensity at approx. 805nm and nanorods/IR800 at approx. 804nm. demonstrated the significant quantum yield enhancement of fluorophore IR800 from approx. 7% to approx. 86% when conjugated with nanoshells and approx. 74% increase with nanorods [52]. This larger difference in the fluorescent enhancement between nanoshells and nanorods is due to large scattering efficiency of nanoshells compared to nanorods [60-61]. This enhanced fluorescence intensity of fluorophores can be used in PTT/PDT application among other applications such as fluorescence imaging and molecule detection.

Nanorods

Gold nanorods are more efficient absorbing and scattering NIR light and thus most promising imaging and photothermal agent compared to gold nanospheres and gold nanoshells [45]. The Gold nanorods are usually synthesized using CTAB which gives positive charge on the surface of the gold nanorods [1-6, 40-50]. In order to remove this positive charge as it is unstable these gold nanorods are coated with poly(styrenesulfonate)which gives negative charge to the surface of gold nanorod and thus attaches to antibody to improve stability, low toxicity of CTAB and increase specificity [1]. Several research demonstrated the significance of nanorod size in photothermal therapy both experimentally and theoretically and found that compared to a smaller size (approx. 17 × 5nm) and larger size (approx. 38 × 11nm), the size approx. 28 × 8nm is ideal for NIR photothermal therapy [1-6, 64]. By tuning the length of the gold nanorods, the longitudinal SPR can be tuned to obtain desired λ NIR. The nanosphere size is mostly determined as diameter nm while the Gold nanorods size is determined as aspect ratio [1-6, 60-70]. Aspect ratio is the ratio of length/width ie. long axis over short axis. As the aspect ratio increases the longitudinal SPR shifts to NIR. It has been found that gold nanorods longitudinal SPR peak position is sensitive compared to transverse SPR and can be influenced with the biomolecules attached to the gold nanorods [12]. The same group also found that the gold nanorods are more specifically located within the cell or on the cell membrane depending on the biomolecule attached to the gold nanorods. Recently, research with gold nanorods on cells and their intake by cells, stability and cytotoxicity are gaining huge attention to understand and design gold nanorods as successful photothermal agents [1-6, 60-70]. The gold nanorods were found to be aggregated in both serum free medium and serum media. The gold nanorods surface functionalized with biomolecules can enhance the stability of gold nanorods in the cells and enhanced gold nanoparticles concentration in cells [1-6, 60-70].

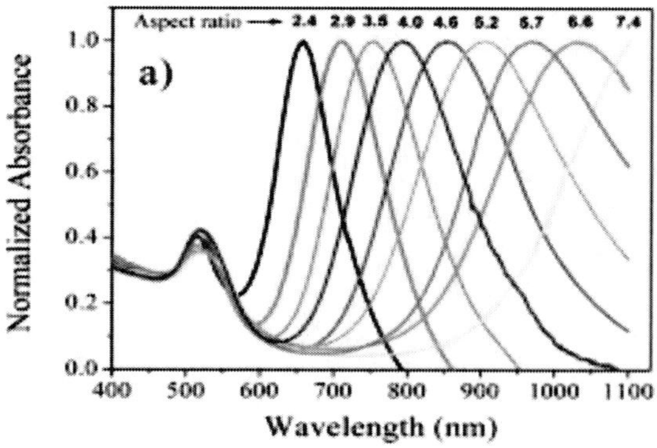

Figure 6. Absorbance spectra of Gold nanorods with varying size and aspect ratio 2.4 to 7.4. Adapted with permission from *Acc. Chem. Res.*, 2008, 41, 1578-1586 [5].

In gold nanorods, the emission and scattering efficiency is related to the aspect ratio. It was found that the longitudinal SPR is directly proportional to aspect ratio [13]. As the length of the gold nanorod increases, the aspect ratio increases and the wavelength shifts to NIR with higher quantum efficiency [13]. In Figure 6, it is shown that the gold nanorods size increased, wavelength shifts from visible to NIR (650-900nm). These gold nanorods which absorbs at NIR wavelength is greatly helpful for deeper cancer cells treatment and this wavelength is not absorbed by the surrounding environment [1-6, 13]. Similar to gold nanoparticles, as gold nanorods aspect ratio varies, absorbance and the color of the scattering light varies as shown in Figure 7 [5]. In Figure 8 that as the size and shape of the nanoparticles (gold nanoparticles and gold nanorods) varies, the wavelength absorbance also varies [5-6]. Gold Nanorods have longer wavelength absorbance compared to gold nanospheres and within gold nanorods, large aspect ratio has longer NIR wavelength absorbance [5-6].

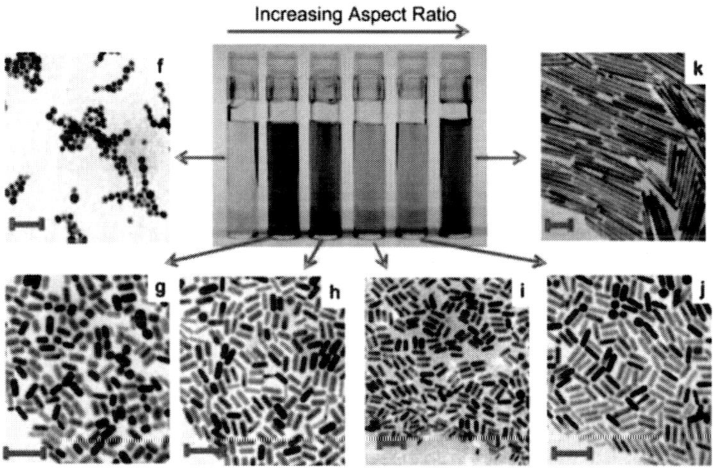

Figure 7. TEM and colors of the Gold nanorod aspect ratio from 1.3 to 5 and 20 for the TEM. Adapted with permission from *Acc. Chem. Res.* 2008, 41, 1721-1730 [6].

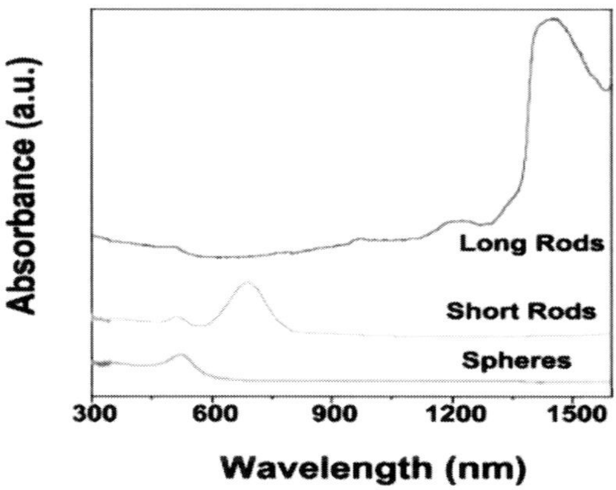

Figure 8. Absorbance of the gold nanoparticles size 8nm, Gold nanorods aspect ratio 3 and 20. Adapted with permission from *Acc. Chem. Res.* 2008, 41, 1721-1730 [6].

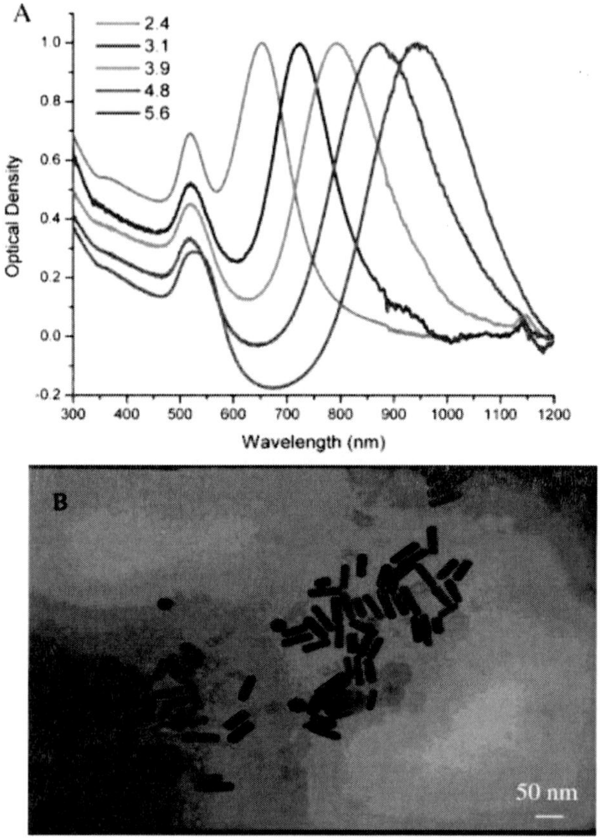

Figure 9. (a) UV-Vis absorption spectrum of GNRs different aspect ratio and (b) TEM image of obtained GNR aspect ratio 3.9. Adapted with permission from *J. Am. Chem. Soc.* 2006, 128, 2115-2120 [6].

El Sayed et al. group demonstrated *in vitro* efficient and highly selective gold nanorods as PTT agent [1, 3-5]. The group demonstrated that by increasing the concentration of gold nanorods within the malignant cells we can use only half the energy of NIR 800nm light [1, 3-5]. Besides passive accumulation, to achieve tumor specific targeting and active accumulation is obtained with using a ligand as demonstrated by El Sayed [1, 3-5]. They used anti-epithelial growth factor receptor antibody (anti EGFR antibody) to achieve active targeting of overexpressed EGFR epithelial carcinoma cell lines (HSC 313 and HOC 3). The concentration of gold nanorods within the cells can be enhanced by attaching to anti EGFR antibody which can specifically bind to malignant cancer cells [1, 3-5]. Thus higher concentration of anti EGFR gold nanorods with in the cell, higher heat is generated from the gold nanorods with NIR light even at low energy thus acts as efficient PTT agent with less damage to surrounding tissue [1, 3-6]. It was found that among different aspect ratio of prepared nanorod in Figure 9(a), the most efficient aspect ratio of gold nanorod 3.9 which has an absorption λ_{max} at which the surrounding tissue absorption is low [1, 14] and Figure 9(b) shows TEM image of gold nanorods with aspect ratio 3.9.

When a cancer cell with gold nanoparticles is irradiated with light at an intensity that can cause ionization, this process not only generates heat that can be applied for PTT and also will generate thermally induced oxidative stress [1, 3-5]. This oxidative stress develops reactive oxygen species like H_2O_2 at high concentrations that are cytotoxic and can lead to irreversible damage to both normal and cancer cell as shown in Figure 12 [1, 3-5, 10-16]. So during NIR irradiation using nanoparticle attached to a biomolecule and irradiation at a light intensity that is below the irradiation threshold can increase the specificity and decrease reactive oxygen species (ROS)concentration, respectively [1, 3-5, 10-16]. These reactive oxygen species (ROS) can be removed during photothermal therapy by using reactive oxygen species scavengers that has antioxidant property [1, 3-5, 10-16]. One such potentreactive oxygen species (ROS) scavenger is Platinum. Introducing PEG platinum coated gold nanorod on HSC3 cancer cells and irradiated with 808nm NIR showed significant cell damage through hyperthermia and inhibition of reactive oxygen species (ROS)cytotoxic effects during PTT [1, 3-5, 10-16]. The photothermal property of the platinum is increased with concentration. The concentration variation is obtained by changing the thickness of platinum coated on the tip of gold nanorod which is further stabilized with PEG. By varying the concentration of platinum, the gold nanorod are modified to thin shell platinum coated gold nanorod (Pt13 Au87NR) and thick shell platinum coated gold nanorod (Pt20Au80NR), Figure 10 and 11 [15-16]. The PTT and reactive oxygen species (ROS) scavenging ability was tested for gold nanorod without platinum, Pt13Au87NR and Pt20Au80NR [15-16]. It was also demonstrated that concentration of nanorod and overlap between the nanorod longitudinal SPR peak and the NIR irradiation light shows significant effect on the localized heat generated [15-16]. Furthermore, it was found that after exposure to NIR light, gold nanorod without platinum generates significant amount of reactive oxygen species and Pt20Au80NR has greater reactive oxygen species (ROS) scavenging property and unbalances the intracellular reactive oxygen species (ROS) levels which is beneficial for cell process compared to Pt13Au87NR [15-16]. The results obtained suggested that nano-Pt system (Pt13Au87NR) acts as more efficient PTT with greater heating efficiency and an efficient reactive oxygen species (ROS) scavenger that prevents harm to normal cells whereas gold nanorod produces significant amount of reactive oxygen species (ROS) after exposed to NIR light [15, 16].

As we already know, PTT can be as an effective treatment by itself and also be combined with conventional chemotherapies to obtain synergistic effect. Huang et al., integrated versatile functions for effective cancer therapy and synthesized a nanocomplex that has simultaneous dual functionality of PDT and PTT using a single NIR 808nm irradiation at low energy [1, 4, 18].

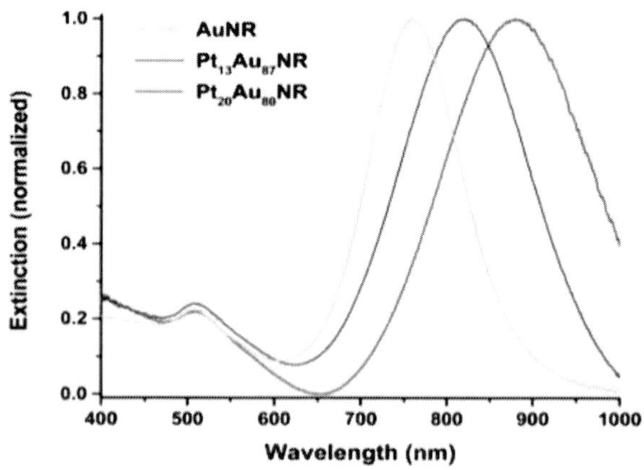

Figure 10. UV Vis absorbance of gold nanorod, thin shell platinum coated gold nanorod (Pt13Au87NR) and thick shell platinum coated gold nanorod (Pt20Au80NR). Adapted with permission from *ACS Nano*, 2017, 11, 579-586 [16].

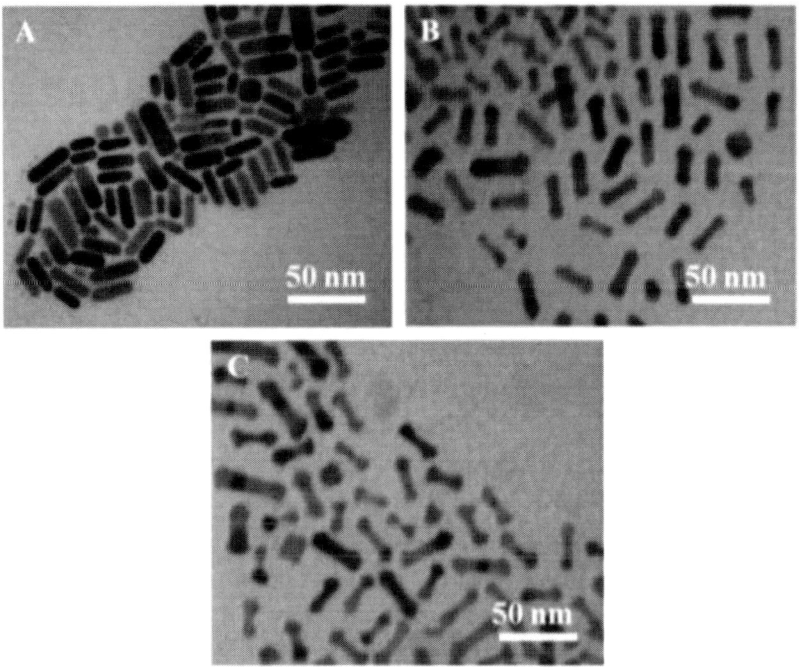

Figure 11. TEM image of (A) Gold nanorod, (B) thin shell platinum coated gold nanorod (Pt13Au87NR) and (c) thick shell platinum coated gold nanorod (Pt20Au80NR). Adapted with permission from *ACS Nano*, 2017, 11, 579-586 [16].

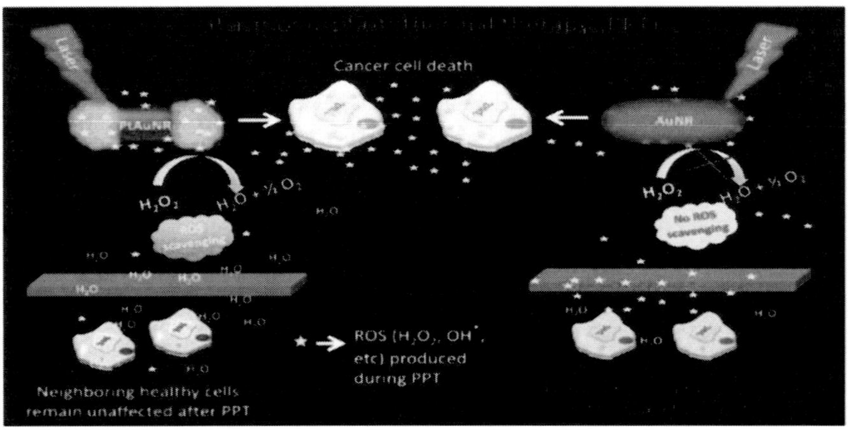

Figure 12. Schematic representation of gold nanorod and platinum coated gold nanorod in PTT. Adapted with permission from *ACS Nano*, 2017, 11, 579-586 [16].

The PTT has different mechanism than PDT. PTT mechanism is explained earlier in this chapter while in PDT, a photosensitizer is irradiated by a specific λ and gives ROS. This ROS is cytotoxic and induces irreversible cell damage [1, 4]. The nanocomplex has a gold nanorods (λ abs = 795nm) that is attached to Photosensitizer, silicon 2,3-naphthalocyanine dihydroxide (SiNC, λ abs = approx. 780nm). It is to be noted that SiNC is more preferable than the widely known Phthalocyanine based molecules. The SiNC-gold nanorod is further coated with PEG alkyl chain at different lengths 1) 6-mercaptohexadecanoic acid (MHA)-PEG-SiNC/gold nanorod, 2) 11-mercaptoundecanoic acid (MUA)-PEG-SiNC/gold nanorod and 3) 16- mercaptohexadecanoic acid (MHDA)-PEG-SiNC/gold nanorod [1, 4, 18]. It is easy to prepare, water soluble, stable and biocompatible. The PEG alkyl chain lengths coated on to the surface of SiNC loaded gold nanorod influences the stability, amount of SiNC on to the gold nanorod and the PDT/PTT combined therapy [1, 4, 18]. It was experimentally showed that Longer the alkyl chain, higher the amount of SiNC and thus enhances effectiveness of PDT/PTT combined therapy [1, 4, 18]. Among the different PEG coated, MHDA-PEG-SiNC/gold nanorod showed higher cell damage efficiency than the remaining two in displaying higher loading of SiNC and SiNC release [18].

Recently, a multifunctional nanosystem was developed that allows fluorescence imaging, PDT and PTT [10-20]. The nanosystem is synthesized with m-PEG-SH coated on gold nanorod, further coated with RRLAC peptide to give gold nanorod-mPEG-SH-RRLAC complex [19]. To this the photosensitizer $AlPcS_4$ (Al (III) phthalocyanine chloride tetrasulfonic acid) is added to give gold nanorod- $AlPcS_4$ complex [19]. The uniqueness of this nanosystem is that NIR 810nm irradiation stimulates the PS release from gold nanorod and NIR around 670nm stimulates PDT. Figure 13(a) shows the UV-Vis spectra peaks at around 670nm for the photosensitizer free $AlPcS_4$ and the peak around 800nm for the gold nanorod- $AlPcS_4$ complex which significantly overlaps with the emission peak of the free $AlPcS_4$ [19]. Here the Photosensitizer acts as a fluorophore in addition to generating ROS. Photosensitizer on gold nanorod is converted from non-fluorescent and non phototoxic to fluorescent and phototoxic with heat generated when irradiated with NIR [19]. The nanocomplex has significant advantages higher cellular uptake, less serum binding, higher bioavailability, stability and low energy is enough for *in vitro* NIR hyperthermia [19]. The gold nanorod- $AlPcS_4$ complex when irradiated with NIR 810nm yields high temperature compared to the control PBS in Figure 13(b). In Figure 13(c) gold nanorod- $AlPcS_4$ complex shows greater reduction in tumor volume compared to PDT or PTT [19].

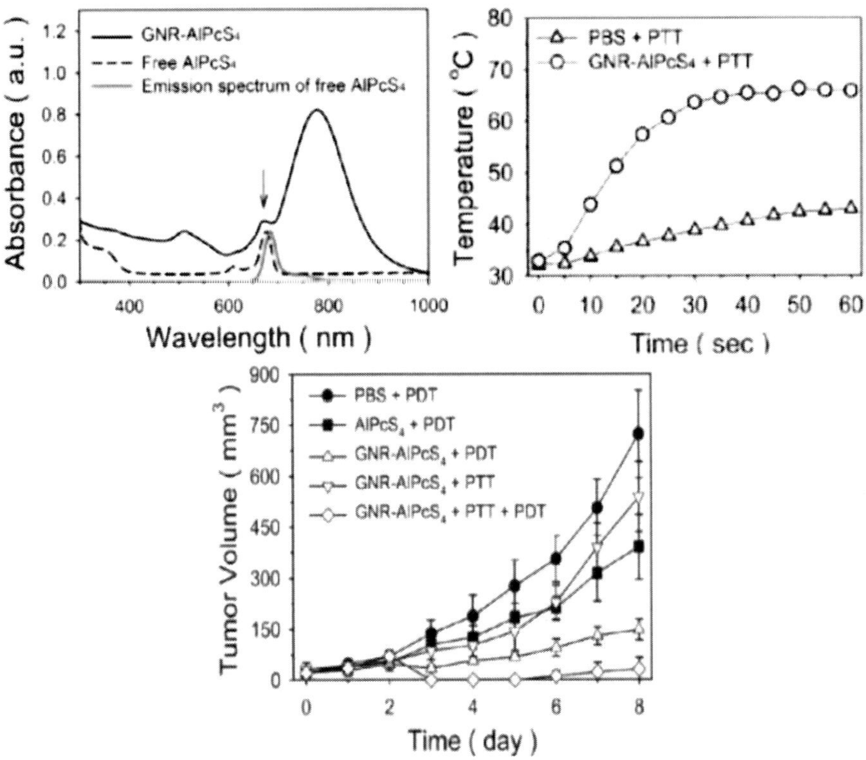

Figure 13. (a) UV- Vis absorption spectra of gold nanorod-$AlPcS_4$ nanocomplex, free $AlPcS_4$ and emission spectra of free $AlPcS_4$. (b) Temperature- time graph of photothermal therapy with control PBS and the gold nanorod-$AlPcS_4$ nanocomplex and (c) Tumor volume after PDT or PTT or both PDT/PTT. Adapted with permission from *ACS Nano,* 2011, 5, 1086-1094 [19].

Inorganic Nanophotothermal Agents

Bismuth Sulfide (Bi_2S_3) Nanoparticles

Zhang et al. synthesized the Aggregation Induced Emission (AIE) dye based BSA coated Bismuth sulfide (Bi_2S_3) nanoparticles which represents efficient, photostable and easy to synthesize nanoprobes for fluorescence imaging and PTT [20-22]. BSA coating on the surface Bi_2S_3 nanoparticles gives remarkable stability and biocompatibility and supports the AIE dye based fluorescence of these nanoparticles. The Hep G2 cells were incubated with these nanoparticles and found bright green fluorescence at 405nm excitation and at 808nm irradiation shows excellent PTT [20-22]. The multifunctionality of the BSA coated bismuth sulfide nanosystems were not only limited to PTT but due to their excellent NIR at 808nm and X-ray absorption can be applied to photoacoustic imaging at 680nm excitation, Radiotherapy and computed tomography imaging, respectively. Bi_2S_3 nanoparticles are more effective as computed tomography contrast agents [20]. This multifunctionality of BSA coated Bi_2S_3 nanoparticles was demonstrated by Wang et al. [21-22]. The results demonstrated Bi_2S_3 nanoparticles can complete removal of tumor in mice with around 100% survival rate of mice after 40 days treatment. Bi_2S_3 nanoparticles with NIR and X-ray irradiation are also remove aggressive metastasis. These excellent properties of Bi_2S_3 nanoparticles show great potential in future real time application in patients [21, 22].

Copper Sulfide (CuS) Nanoparticles

Positron emission tomography (PET) is a sensitive technique that requires the radioactive atoms to detect the cellular changes at pico molar concentrations [64-66]. The radioactive element is conjugated with nanoparticles using chelators to form stable complex. The nanotechnology integrated PET technique use chelate to give precise *in vivo* imaging [64-

66]. The integrated system has limitations such as exposure to radiation and trasnschelation. Zhou et al. were the first to contribute to improve the PET imaging technology in terms of chelator free Nanoparticles by incorporating [^{64}Cu] into CuS nanoparticles [66]. ^{64}Cu is a known radioactive element in PET imaging. Extensive research was conducted to prepare and modify Copper Sulfide (CuS) nanoparticles which absorb NIR light at 808nm and can convert to heat making them potential in the cancer photothermal therapy [66]. The [^{64}Cu] CuS nanoparticles were prepared as an efficient image guide photothermal therapy. These CuS nanoparticles were stabilized with thioglycolic acid or citrate or PEG. Among these PEG coated [^{64}Cu] CuS nanoparticles were realized to be more stable [66]. It was demonstrated that PEG coated [^{64}Cu] CuS nanoparticles have enhanced absorption coefficient approx. $8.66 \times 10^7 M^{-1}cm^{-1}$, biocompatible, longer circulation time, low power laser is enough and enhanced bioavailability and when given intratumorally to U87 tumor, these nanoparticles reaches tumor through passive process and irradiated with 808nm was found to increase tumor temperature in 5 mins and causes cancer cell necrosis thus completely eliminate the tumor [66]. The *in vivo* PET imaging was performed 24 hrs after injection of [^{64}Cu] CuS nanoparticles results in clear visualization of tumor [66].

Silica Nanoparticles

Porous silica nanoparticles were widely explored in chemo-photothermal therapy conjugated a dye, as a hybrid with copolymers such as polyaniline and in multidrug resistant tumor because of excellent surface functionalization, high loading capability, ease of release, biocompatibility and biodistribution [67-68]. Polyaniline is covalently attached to the Porous silica nanoparticle to give polyaniline/porous silicon hybrid nanoparticle and further loaded with chemotherapeutic agent [67-68]. These nanoparticles were intratumorally administered to 4T1 tumor and irradiated with NIR laser for 10 mins which increases the temperature of tumor to 44-45°C and shows remarkable decrease in tumor growth

(90.6%). Tumor acidic pH 5.0 stimulates the drug release and the NIR irradiation further increases the drug release by 17.7%. The NIR irradiation also allows the increased cell permeability which facilitates the drug molecules to reach the microenvironment of cells via endocytosis thus enhances the efficiency of chemotherapy [67-68]. The drug loaded polyaniline porous silica hybrid nanoparticles with NIR demonstrated promising synergistic chemo/photothermal treatment for cancer cells and drug resistant cancer cells [67, 68].

Single Walled and Multi Walled CNT

Gold nanoparticles though hugely considered for cancer therapy. But the gold nanoparticle have strong absorption in the visible region which has less tumor penetration capacity, low NIR absorption and less thermal conductivity unlike carbon nanotubes [20-30]. Carbon nanotubes (CNT) are cylindrical like shape made up of carbon. These are small size and high surface area which enables the carbon nanotubes to easily adsorb or attach to biomolecules [20-30]. This ease of surface functionalization enables its promising uses in drug delivery or chemotherapy. These carbon nanotubes (CNT) can act as dual and synergistic agent with both NIR light and visible light where using NIR light, the carbon nanotubes (CNT) can be used as photothermal agent and using visible light, the carbon nanotubes (CNT) can generate singlet oxygen species to provide PDT [20-30]. There are two types single walled carbon nanotubes (SWNT) and multiwalled carbon nanotubes (MWNT). We are going to discuss few examples in this chapter. The high optical absorption at near infrared region of SWNTs with low background and high sensitivity and photostability are proving to be promising remarkable results in cancer diagnosis imaging [23] and thermal treatment for primary tumor and metastasis tumor [20-30]. With surface modifications, these SWNT can enhance *in vivo* and *in vitro* specificity in photothermal therapy [20-30]. Carbon Nanotubes (CNT) have capability of simultaneous generate luminescence and heat which is by the relaxation of excited π electrons between carbon atoms. The carbon nanotube with large diameter has great heat efficiency [24]. CNT have large surface area, easy surface modification, high drug/molecule loading capacity, stable nanosize and large NIR absorbance and are becoming increasingly popular as PTT agent to absorb NIR light and can cause localized photothermal effect with no significant *in vivo* toxicity [20-30]. Because of easy surface modifications these CNT can be easily conjugated with ligands and utilized for changing the hydrophobic nature of CNT and can be utilized for selectively treating cancer cells [20-30]. Metal complexes mainly the transition metal complexes have huge attention in cancer research as some of these complexes may mimic the compounds/ molecules to provide anticancer efficacy for example cisplatin [20-30]. And cancer cells become resistant to such treatments and demanding for more research in anticancer agents. As an alternative to these cisplatin, Ruthenium (II) complexes are gaining attention as anticancer agent. Ruthenium (II) polypyridyl complexes (Figure 14) have strong absorbing and luminescence properties and when irradiated with UV or visible light (Figure 15) can generate singlet oxygen species and is an efficient PDT agent towards cancer cells [30]. The SWCNT have pores such that the Ruthenium (II) polypyridyl complex forms non covalent π-π interactions [30]. These Ruthenium (II) SWCNT complexes when irradiated with 808nm at a low laser power shows both PDT and PTT effect thus results in synergistic anticancer efficacy [30]. The Ruthenium (II) complex enhanced the hydrophilicity and stability of hydrophobic SWCNT for 3 months [30]. They have more photothermal conversion efficiency (P.C.E.) than gold nanorods and CuS nanoparticles. Ruthenium (II) SWCNT nanoparticles were >96% non cytotoxic and when injected to the intratumorally and irradiated with NIR laser 808nm shows that the tumor temperature is increased in 5 mins necrosis and tumor size is decreased over time [30].

Zhou et al. prepared SWNT shows narrow absorption peak around 980nm. *In vivo* and *in vitro* SWNT was attached to folate antibody which has more specificity to cancer cells so minimal damage to healthy cells [25]. It was found that compared to SWNT, folate SWNT when irradiated with around 980nm NIR light shows higher photocytotoxcity and selectivity [25]. Another group showed that SWNT were more efficient theranostic agent in regards to dose of photothermal agent and NIR power energy compared with widely used gold nanorod 22-24. They demonstrated that the SWNT-PEG has high intracellularization, an effective imaging agent and complete tumor damage using 808nm at low power energy [22-24]. To understand the efficiency of image guided PTT over conventional cancer surgery, Liang et al. injected PEG-SWNT at the site of primary tumor, within 90 minutes, the PEG-SWNT reaches to the secondary tumor site [27]. After that, *In vivo* imaging and PTT therapy is performed using a low dose of 0.2mg/Kg of PEG-SWNT [27]. The SWNT has intrinsic NIR photoluminescence. This helps in imaging exact location of tumor and the possible pathways for cancer cells to metastasize from primary tumor to secondary tumor site, i.e., lymph node. In PTT, the primary tumor and secondary tumor sites were irradiated at low dose and low energy of 0.5w/cm2 and 0.8w/cm2, respectively where the temperature reaches >40°C [27]. The results of cancer therapy through PTT to both primary and secondary tumor sites, PTT to primary tumor and surgery indicated that PTT to both primary and secondary tumor sites has enhanced survival rate after treatment with no signs of tumor [27]. Thus demonstrates the efficiency of PTT on both primary tumor and metastasis over cancer surgery [26, 27]. Recently a group found a way to treat single cancer cells or their clusters at a high temperature around 95°C at a short exposure time using long laser wavelength [28]. Short time exposure to laser so less heat dissipation to surrounding healthy cells so less damage to healthy cells while effective cancer cell damage. The results are found to be an effective way to photothermal therapy of metastatic cancer and as wells as single cells [28]. Carbon nanotubes can do just more than PTT and imaging. They are beneficial in drug delivery 29 and specific targeting using biomolecules. For example, MWCNT can be used as drug carrier for various drugs like Epirubicin [31]. Another research found in both *in vitro* and *in vivo* studies that MWCNT surface has been functionalized with the drug 10-hydroxycamptothecin (HCPT) is improved the efficiency of drug cancer activity and low toxicity. PEG-MWCNT can be used as an efficient drug delivery for multiresistance drug problem in chemotherapy [32]. The heat generated around 42°C for 10s NIR irradiation of MWCNT increased the cell damage and cell permeability for drugs like oxaliplatin or mitomycin C. Thus show excellent chemophototherapeutics of MWCNT [33]. MWCNTs themselves can also inhibit tumor growth and also can treat highly metastatic and chemotherapy resistant cancer [69].

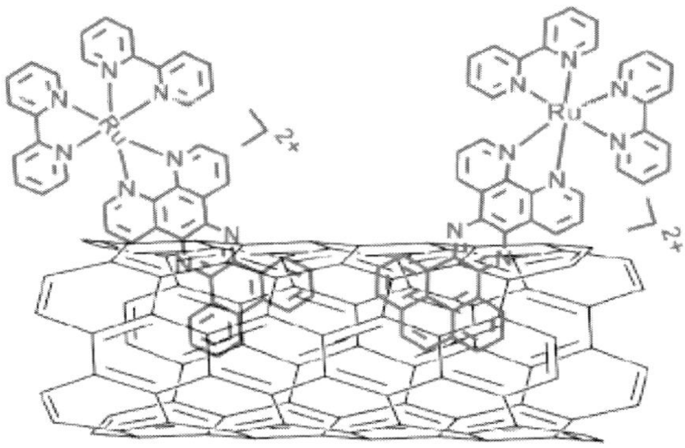

Figure 14. Schematic representation of Ru(II)SWCNT complex. Adapted with permission from *ACS Appl. Mater. Interfaces*, 2015, 7, 23278-23290 [30].

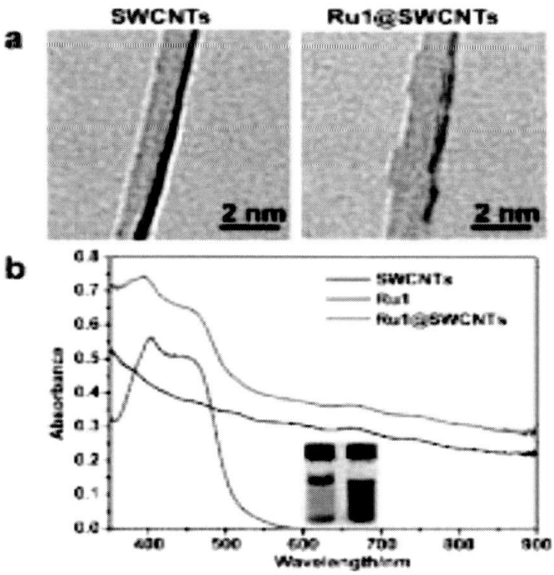

Figure 15. (a) TEM images of the Ru(II)SWCNTs. b)UV/Vis spectra of Ru(II)SWCNT, Ru(II) and SWCNT in aqueous solution. Adapted with permission from *ACS Appl. Mater. Interfaces*, 2015, 7, 23278-23290 [30].

Compared to SWNT, MWNT have large absorption spectra, large surface area and more electrons available per MWNT particle which absorbs greater amount of NIR energy [34-36]. Thus, MWNT are more efficient to generate greater amounts of heat [34, 35, 36]. MWCNT can be functionalized to be cancer cell receptor specific by non covalently conjugated with HSA and thus enhances the selectivity and specificity and *in vitro* when irradiated at 808nm 2 mins demonstrated that cancer cells were necrotic [70]. Also another group showed both *in vitro* and *in vivo* efficiency of MWCNT with no systemic toxicity complete tumor removal when irradiated with NIR 1064nm provides complete elimination of tumor and treated mice were tumor free for more than 3.5 months in 80% of treated mice. MWCNT less amount is enough to absorb more NIR due to more electrons [71]. MWCNT were also been study for their efficacy towards cancer therapy A study was conducted to demonstrate the high heating efficiency of MWCNT on prostate cancer cells and renal cancer cells with laser irradiation at 1064nm with irradiance at 15.3W/cm^2 [37]. The temperature elevated to greater amounts in presence of MWCNT compared to without MWCNT. The efficacy of this treatment is measured with cytoxicity and heat shock protein (HSP) expression [37]. HSP expression can determine the recurrence of cancer. So diminishing the HSP expression is required. MWCNT with laser resulted decreased the HSP expression and enhanced cytoxicity [37]. To demonstrate the *in vivo* and *in vitro* efficacy of MWCNT towards metastatic cancer, PEG-MWCNT irradiated at 808nm can effectively suppress the tumor growth in bone metastasis with increasing dose of MWCNT [34]. Highly stable and hydrophilic PEG-MWCNT are highly efficient around 808nm NIR light to decrease the tumor volume [38]. Overexpression of HSA can internalize large amounts of albumin has been well researched. Thus attaching MWCNT to HSA can be useful in selectively target cancer cells with high HSA levels on the cell membrane and can significantly with 808nm NIR. This is similar to attaching folate on the CNT for folate receptors on the cancer cell membrane is discussed in this chapter. In MWCNT-HSA, the HSA protein binds to chemodrugs and thus acts as multifunctional nanosystem [39].

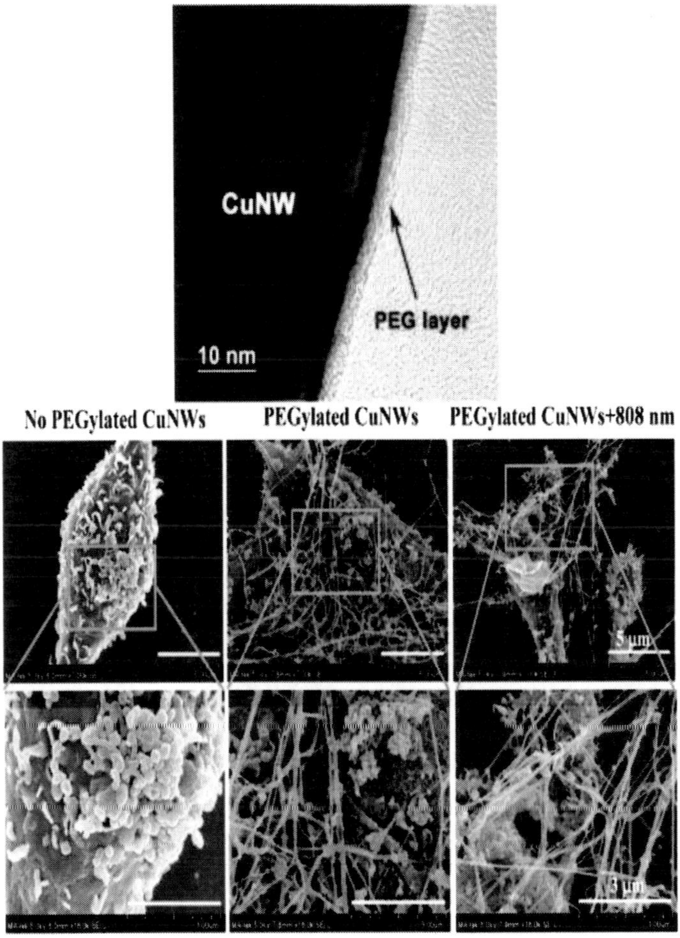

Figure 16. (a) TEM image of hydrophilic PEGylated CuNW, (b) shows SEM image of tumor cells with and without PEGylated CuNW and with and without Irradiation at 808nm. Adapted with permission from *ACS Appl. Mater. Interfaces*, 2016, 8, 12082-12090 [2].

Metal Nanowires

Due to unique properties and low toxicity of noble metal like gold nanoparticles and silver nanoparticles, they have been widely explored in cancer imaging and photothermal therapy [1-5, 10, 40]. There are also new photothermal agents that are emerging with excellent heating properties.

Copper Nanowires (CuNW)

With development of different Metal nanowires in PTT application, first time it was showed that PEGylated CuNW can be used as photothermal agent [2]. In synthesis of PEGylated CuNW, initially Copper chloride (CuCl) is dissolved in oleylamine and heated to 110°C for 50 mins followed by secondary heating again at 250°C for another 50mins[2]. The resulted CuNW are hydrophobic. In order to enhance the aqueous dispersibility, these hydrophobic CuNW are attached to PEG and heated at 50°C for 4hrs to obtain water soluble PEGylated CuNW as shown in Figure 16a [2]. To understand the photothermal heating property, PEGylated CuNW at a concentration of 2.5mg/ml was injected into a tumor volume approx. 50-60mm^3 and incubated for 1 hour followed by irradiation with NIR 808nm[2]. At 6 mins exposure of the NIR, the tumor temperature was raised > 50°C leading to irreversible cell damage and further observed significant reduction in tumor size [2]. The SEM image in Figure 16b shows the potential effectiveness of photothermal effect of the prepared PEGylated CuNWon tumor cells with NIR irradiation at 808nm compared to PEGylated CuNW on tumor cells without NIR irradiation [2]. The no PEGylated CuNW tumor cell images are shown as a control. These results demonstrated that PEGylated CuNW or NIR laser irradiation alone had little or no effect on cell viability [2].

Tellurium Platinum Nanowires (Te Pt NW)

A tumor of volume 200mm^3 was treated with aqueous suspension of Te Pt Nanowires with effective concentration of 10ug/ml and irradiated with 808nm for 10-12 mins [40]. After 18 days of irradiation, the tumor size was measured and it was found that the tumor size was reduced to approx. 95 mm^3. Te Pt NWrs were synthesized using tellurium dioxide (TeO_2) and hydrazine to give Tellurium nanowires further treated with CTAB and $PtCl_2$ at 100°C for 12 hrs to give Te Pt Nanowires [40]. These nanowires are effective photothermal agents with high absorbance at NIR and are photostable. These nanowires when irradiated with 808nm can increase the temperature of tumor to a high as approx. 62°C and results in irreversible cell damage [40].

The gold nanoparticle (AuNP) has absorption around approx. 530nm while such absorption is not suitable for deeper penetration cancer tissue treatment and diagnosis. Hence, there are ways to apply AuNP for NIR photothermal activity at deeper tissues by modifying their surface. Yao He et al. [10] demonstrated experimentally a remarkable highly efficient photothermal agent that can increase the temp to high as approx. 80°C within short period approx. 3mins in carcinoma cells, A549 cells and Hela cells by attaching AuNP on the surface of SiNW [40]. This high increase is due to synergistic effect of AuNP and SiNW. The SiNW absorb around NIR 808nm to heat the surrounding tissue while AuNP are on the surface of the SiNW can increase this temperature. These AuNP SiNW were further PEGylated to decrease the cytotoxicity and enhance bio-compatibility. PEG-AuNP-SiNW tested for cytotoxicity at concentrations of 75ug/ml, 150ug/ml and 300ug/ml and showed the results that PEG-AuNP-SiNW itself is less toxic [40]. Only with NIR irradiation for 3 mins. shows high efficiency photothermal activity that all carcinoma cells were destructed.

Polymer Nanoparticles

Recently Yang et al. [41]developed a biosafe and biodegradable protein coated coordination polymer nanoparticle using BSA protein, Gallic acid and Fe (III) chloride hexahydrate to give BSA-GA-Fe nanoparticles [42-44]. These nanoparticles have proved to be excellent photoacoustic agents and photothermal agents with strong optical absorption at visible and NIR region thus can be applied as real time imaging guided photothermal therapy [42-44]. The average diameter of the nanoparticles is around 3.5nm which helps in easy removal from the body. They are homogenous distribution in water, highly photostable, low cytoxicity and high biocompatible [42-44]. After 24hrs intravenous injection of these nanoparticles to the tumor site, the *in vivo* results show that there is a negligible amount of nanoparticles retained in the mice [41-44]. Thus showing the ultra small size of nanoparticles is easily removed from mice. These nanoparticles are dose dependent hyperthermic [42-44]. As the concentration increased from 0.03mM to 1.00 mM, temperature of the nanoparticles solution also increased. The tumor volume is completely reduced when irradiated at 808nm with these nanoparticles.

CuS nanoparticles have been extensively studied because of the broad absorbance near infrared region and excellent thermal capacity to induce PTT thus CuS nanoparticle acts as an excellent tumor diagnosis and therapy [70-73]. But CuS nanoparticles also have limitations of solubility and non specific interactions. And several dyes have excellent fluorescent capacity. It is well-known that multifunctional nanoparticles are in more demand currently [70-73]. But when CuS and dye combine lowers the function, i.e., CuS quenches the fluorescence capacity of the dye. Also CuS has poor tumor site specificity. So in order to combine these functionalities and generate a multifunctional polymer nanoparticle, another group developed enzyme responsive photothermal therapy by using hyaluronic acid nanoparticle (HANP) in SCC7 tumor [70-73]. For this purpose, they explored the idea of the presence of excess levels of enzyme Hyaluronidase, which is a significant biomarker to confirm the tumor-igenesis and metastasis in patient [72]. For imaging purpose, this nanoparticles is labeled with near infrared dye such as Cy5.5 and for photoacoustic (PA) imaging and PTT, these nanoparticles are loaded with CuS. Thus results in Cy5.5 HANP/CuS, i.e., Cy5.5/HANPC nanosystem Figure 17 which is an inactive form [72]. These HANPC has dual targeting property, highly soluble in variety of solvents, high UV/Vis/NIR absorbance and greater cell targeting efficiency. When cy5.5/HANPC nanoparticles accumulated in the tumor via endocytosis, the Hyaluronidase enzyme inside the tumor activates the nanoparticles by degrades the nanoparticles into Cy5.5/HANP and CuS. The cy5.5/HANP give a strong fluorescence signal. CuS shows PA imaging [72-73]. When given IV, after 6 hrs of post introduction of the nanoparticles, the CuS in the tumor on irradiated with NIR 808nm for 10 mins showed that significant tumor suppression [73] as shown in Figure 17.

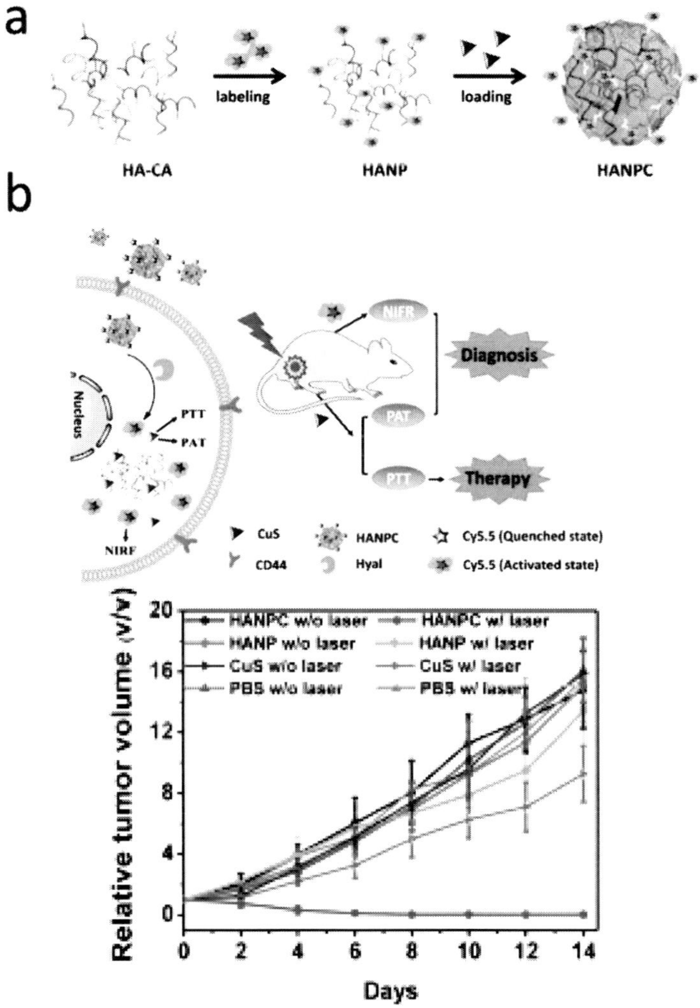

Figure 17. Tumor growth rate in the following groups: PBS group with and without laser irradiation free CuS group with and without laser irradiation. HANP group with and without laser irradiation and HANPC group with and without laser irradiation. Also we can observe significant tumor volume reduction in HANPC treated tumor group. Adapted with permission from *ACS Nano*, 2014, 8, 12250-12258 [73].

Most research prefers multifunctional polymer nanoparticles over single modal nanoparticle. Among several multifunctionality, combining PDT and PTT activity into a single nanoparticle is not uncommon [70-75]. To achieve this goal, most common PDT agents used are photosensitizers in combination with an inorganic PTT agent [70-75]. Such nanoparticles are not without disadvantages such as low ROS production, non specific activity of antibody attached photosensitizer, debatable excretion and toxicity of inorganic PTT agents [70-75]. To overcome such disadvantages a group developed a strategy to synthesize similar type of material to do multiple functions, i.e., two polymers, one polymer acts as PDT agent and second polymer acts as PTT agent. The polymer 1: poly[9,9-bis(2-(2-(2-methoxyethoxy)-ethoxy)-ethyl)fluorenyldivinylene]-alt-4,7-(2,1,3-benzothiadiazole) and polymer 2: poly[(4,4,9,9-tetrakis(4-(octyloxy)-phenyl)-4,9-dihydro-s-indacenol-dithiophene-2,7-diyl)-alt-co-4,9-bis(thiophen-2-yl)-6,7-bis(4-(hexyloxy)phenyl)-thiadiazolo-quinoxaline] are co-loaded and encapsulated in lipid-polyethylene glycol to give a small size approx. 30nm multifunctional conjugated polymer nanoparticle [74]. The polymer 1 acts as a photosensitizer, PDT agent. The polymer 2 shows photothermal activity due to high absorption coefficient in NIR region (600nm to 1100nm). These conjugated polymer nanoparticles (CPN) when intravenously injected into mice containing SKBR3 tumor and irradiated with white light for 1hr followed by 808nm irradiation for 6 mins shows efficient ROS production, approx. 47.6% photothermal conversion efficiency and tumor temperature increased to approx. 60°C followed by tumor necrosis and unaffected normal tissue [74]. The study was continued for 16 days and observed that tumor growth was reduced to approx. 47% [74].

Human serum albumin nanoparticles were developed which encapsulates the hydrophobic toxic IR780 dye which has excellent NIR fluorescent imaging/PDT/PTT property [75-77, 93, 94]. The HSA enhances the solubility and drug tolerance to 1000 fold and 2.5mg/kg to 25mg/kg, respectively. When HSA/IR780 nanoparticles are given to a tumor and irradiated with NIR 808nm 1W/cm2 for 10 mins simultaneously generate heat and ROS [75-77, 93, 94]. Also the energy from excited singlet oxygen species could be converted to heat and allows irreversible tumor damage. IR780 with polypyrrole nanoparticles with HA gives similar result and allows specific tumor accumulation by binding to CD44 receptors overexpressed in cancer cells [75-77, 93, 94]. Another NIR dye IR825 is conjugated to HSA to give approx. 8nm HSA/IR825 nanoparticles with better fluorescent quantum yield useful in *in vivo* imaging and 100% photothermal tumor treatment at low dose. These nanoparticles when conjugated with Gadolinium could result in remarkably reduced lymphatic tumor metastasis and enhance prolonged survival [75-77, 93, 94]. Unlike the widely explored PEDOT:PSS, PMAM and porphyrin nanoparticles for photothermal therapy, polypyrrole nano-particles are excellent biocompatible PTT agent without requiring any surface modification like PEG coating [79-81]. The *in vivo* photothermal cancer therapy application of polypyrrole nanoparticles was demonstrated on 4T1 tumor using 808 nm laser with a power density of 1 W cm^2 for 5 min tumor temperature was increased to approx. 60°C with no tumor or tumor regrowth without side effects after treatment [79-81]. The NIR mediated high temperature increase of the tumor of approx. 60°C was resulted from photothermal conversion efficiency approx. 44.7% [79-81].

Using ligands or antibody the tumor specific treatment with nanoparticles is achieved but the unaltered biodistribution and uneven intratumoral distribution can cause reduced efficiency of the treatment with tumor recurrence [75-81]. Alternatively, Zhang et al. demonstrated a nanocomposite without a ligand by using polymer with a reversible thermoresponsive property [81]. A multifunctional thermo and PH responsive polymer poly(N-isopropylacrylamide-co-acrylic acid) PNIPAM, which has electrostatic interactions with drug molecules in a Au@SiO_2nanocomposite [81]. When the tumor temperature rises to higher than 32°C, the PNIPAM polymer hydrophilic is transition to hydrophobic and when temperature decreases the nanoparticle shows reverse structure. This nanocomposite with polymer coating enables enhanced longer circulation time and synergistic efficacy in tumor by both phototherapy and chemotherapy. External stimuli such as heat and NIR laser enables enhanced nanoparticles accumulation within tumor such that tumor can be irradiated whenever necessary within 14 days [81]. This group also demonstrated the antitumor activity by increasing the tumor temperature by varying the laser power at NIR 808nm laser [81]. When nanoparticles reach tumor, the NIR laser shrinks the size of nanoparticles such that the electrostatic interaction of polymer with drug is dissociated and the drug release from the nanoparticles [81]. These nanocomposite *in vivo* can reduce the dosage required for any further tumor laser treatment at surface or deep level, enhance bioavailability and high drug dose loading with less systemic side effects to normal tissues. The PNIPAM Au@SiO_2 nanocomposite is effective for tumor inhibition and also for lung metastasis without necessary for active ligands [81]. Polyaniline nanoparticles are more photostable, has high absorption coefficient at NIR region and an excellent PTT agent compared to most popular PTT agents such as gold nanorods, CuS and CNT [83,8 4, 93]. Although polyaniline nanoparticles are hydrophobic when coated with F-127 polymer the solubility is enhanced to result water soluble monodisperse nanoparticles [83, 84, 93]. In acidic tumor environment, the emeralidine base (EB) form gets converted to the emeralidine salt (ES) form. This ES form has at NIR absorption when λ_{ex} 810nm and these nanoparticles are relatively large and require additional research for polyaniline EB nanoparticle to enhance specific tumor accumulation [93]. Another group added stabilizers to obtain polyaniline nanoparticles small size approx. 48.5 ± 1.5nm shows much higher photothermal conversion efficiency approx. 48.5%. Both *in vitro* and *in vivo* results at 808nm laser for 20 mins tumors were completely eliminated. To improve the specific tumor accumulation these polyaniline nanoparticles are attached to Hyaluronic acid (HA) for CD44 cancer cells [83, 84].

Another class of conducting polymers is donor acceptor conjugated polymers such as poly[9,9-bis(4-(2-ethylhexyl)phenyl)fluorene-alt-co-6,7-bis(4-(hexyloxy)phenyl)-4,9-di(thiophen-2-yl)-thiadiazoloquinoxaline] (PFTTQ) with charge transfer between donor and acceptor has advanced photostable, size fine-tuned and broad absorption 700 to 850 nm [85]. The *in vivo* 808 nm laser for 10 mins shows temperature of tumor to be approx. 45°C and eliminates tumors completely [85]. While applying chemo-therapeutics to the nanomaterials the main goal is to have efficient maximize drug loading capacity and improve the specificity with less undesirable cross linking of the drug or ligands molecules [85]. While ADT therapy is a current advanced treatment for prostate cancer but the prostate cancer cells can be resistant to this treatment leading to castration-resistant PCa a major concern currently. This can also contribute to PCa tumor metastasis where the tumor cells due to EMT can cause cancer recurrence [85]. To address this castration-resistant prostate cancer (PCa) Zhang et al. developed PDA/PAH/Cit/drug loaded nanoparticle demonstrated to be effective to diagnose and treat PCa [86]. Among several different polymers like PEG, polypyrrole, polyaniline, poly-(allylamine hydrochloride)- citraconic anhydride (PAH/Cit) polymer is one such polymer whose based nanoparticles useful in treating tumor and its angiogenesis [86]. PAH/Cit is a pH-responsive charge-reversal polymer in acidic environment can be easily converted back to cationic poly (allylamine) and allows efficient control drug release [86]. They have demonstrated promising results both *in vivo* and *in vitro* to treat PCa cells using approx. 73.3 nm poly dopamine(PDA) nanoparticles that are functionalized with PAH/Cit which promotes dispersibility and mainly its nature as pH-responsive charge-reversal polymer in the acidic environment of tumor (pH approx. 5.0) [78]. Its broad NIR region absorption, pH controlled drug release with approx. 30% drug loading capacity, approx. 40% photothermal conversion efficiency and when tumor size of approx. 100mm^3 irradiated with 808 nm showed no tumor recurrence making it an effective synergistic theranostic agent [86]. The PEGylated polydopamine nanoparticle loaded with a dye and drug demonstrates acidic tumor environment responsive combined therapy PTT/PDT and chemotherapy using low dose therapeutic agent with absorption 600 nm to 1000 nm and reduced side effects [87].

Conclusion

In summary, we have presented a brief review of the photothermal nanoparticles. We presented the mechanism for unique surface plasmon resonance properties for nanoparticles, the heat generated, the efficiency of *In vitro* and *In vivo* photothermal therapy and the efficiency as multifunctional nanoparticles (such as Photothermal therapy and Photodynamic therapy or Photoacoustic imaging). As discussed there are extremely benefits of photothermal therapy and are highly attractive to researchers but further research is required to understand the consequence of long term and short term effects of nanoparticles in photothermal therapy, improvement in the metastasis condition after photothermal therapy in cancer, bioavailability and cytotoxicity of nanoparticles along with absorption, distribution, metabolism and excretion (ADME) photothermal agents. More research on particle size and shape is required to explore the significance in efficient and effective near infrared light absorption. Most of the nanoparticles were tested to see for any morphological changes to the other organs but microenvironment changes should be also considered to have more understanding of potential toxicities which requires more research. Besides the above mentioned and discussed nanosystems, there are advanced cell mediated nanosystems too with excellent hyperthermia properties. These although advanced have limited by their cost and more relatively complicated equipment for synthesis. Also there is diverse cell morphologies which requires more research to explore the cell mediated nanosystems and improve towards photothermal therapy [88].

Acknowledgments

We thanks NSF-PREM grant # DMR-1205194, NSF CREST grant # 1547754 and NSF RISE grant # 1547836 for their generous funding.

REFERENCES

[1] Huang, X.; El-Sayed, I.H.; Qian, W.; El-Sayed, M.A. *J. Am. Chem. Soc.* 2006, 128, 2115-2120.

[2] Li, K.; Chu, H.; Lin, Y.; Tuan, H.; Hu, Y. *ACS Appl. Mater. Interfaces*, 2016, 8, 12082-12090.

[3] Link, S.; El-Sayed, M. A. *Int. Rev.* Phys. Chem., 2000, 19, 409-453.

[4] Huang, X.; El-Sayed, M.A. *J. Adv. Res,* 2010, 1, 13-28.

[5] Jain, P. K.; Huang, X.; El-Sayed, I. H.; El-Sayed, M. A. *Acc. Chem. Res.,* 2008, 41, 1578-1586.

[6] Murphy, C. J.; Gole, A.M.; Stone, A.M.; Sisco, P.N.; Alkilany, A.M.; Goldsmith, E. C.; Baxter, S.C. *Acc. Chem. Res.* 2008, 41, 1721-1730.

[7] *American Cancer Society.* https://www.cancer.org/research/cancer-facts-statistics/all-cancer-facts-figures/cancer-facts-figures-2017.html.

[8] *National Cancer Institute.* https://www.cancer.gov/about-cancer/understanding/statistics..

[9] Hainfeld, J. F.; Lin, L.; Slatkin, D. N.; Dilmanian, F. A.; Vadas, T. M.; Smilowitz, H. M. *Nanomedicine,* 2014, 10, 1609-1617.

[10] Su, Y.; Wei, X.; Peng, F.; Zhong, Y.; Lu, Y.; Su, S.; Xu, T.; Lee, S.; He, Y. *Nano Lett.*, 2012, 12, 1845-1850.

[11] El-Sayed, I. H.; Huang, X.; El-Sayed, M. A. *Cancer Letters,* 2006, 239, 129-35.

[12] Zhang, W.; Ji, Y.; Meng, J.; Wu, X.; Xu, H. *PLoS ONE,* 2012, 7, 1-11.

[13] Fang, Y.; Chang, W.; Willingham, B.; Swanglap; Dominguez-Medina, S.; Link, S. *ACS Nano,* 2012, 6, 7177-7184.

[14] El-Sayed, I. H.; Huang, X.; El-Sayed, M. A. *Nano Lett.,* 2005, 5, 829–834.

[15] Minai, L.; Yeheskely-Hayon, D.; Yelin, D. *Scientific Reports,* 2013, 3, 1-7.

[16] Aioub, M.; Panikkanvalappil, S. R.; El-Sayed, M. A. *ACS Nano*, 2017, 11, 579-586.

[17] Tang, J.; Jiang, X.; Wang, L.; Zhang, H.; Hu, Z.; Liu, Y.; Wu, X.; Chen, C. *Nanoscale*, 2014, 6, 3670-3678.

[18] Bhana, S.; Connor, R. O.; Johnson, J.; Ziebarth, J.D.; Henderson, L.; Huang, X. *J. Colloid Interface Sci.*, 2016, 469, 8-16.

[19] Jang, B.; Park, J. Y.; Tung, C. H.; Kim, I. H.; Choi, Y. *ACS Nano,* 2011, 5,1086-1094.

[20] Rabin, O.; Perez, J. M.; Grimm, J.; Gregory Wojtkiewicz, G.; Weisslede, R. *Nature Materials,* 2006, 5, 118-122.

[21] Wang, Y.; Wu, Y.; Liu, Y.; Shen, J.; Lv, L.; Li, L.; Yang, L.; Zeng, J.; Wang, Y.; Zhang, L. W.; Li, Z.; Gao, M.; Chai, Z. *Adv. Funct. Mater.,* 2016, 26, 5335-5344.

[22] Wang, K.; Zhuang, J.; Chen, L.; Xu, D.; Zhang, X.; Chen, Z.; Wei, Y, Zhang, Y. *Colloids Surf B Biointerfaces,* 2017, 160, 297-304.

[23] Welsher, K.; Liu, Z.; Sherlock, S. P.; Robinson, J. T.; Chen, Z.; Daranciang, D. Dai, H. *Nature Nanotechnology,* 2009, 4, 773-780.

[24] Siregar, S.; Haq, I, U.; Nagaoka, R.; Saijo, Y. arXiv: 1611.08094 [*Physics Med-ph*], 2016,1, 1-12.

[25] Zhou, F.; Xing, D.; Ou, Z.; Wu, B.; Resasco, D. E.; Chen, W. R. *J Biomed Opt.,* 2009, 14, 1-7.

[26] Robinson, J. T.; Welsher, K.; Tabakman, S. M.; Sherlock, S. P.; Wang, H.; Luong, R.; Dai, H. *Nano Res.,* 2010, 3, 779-793.

[27] Liang, C.; Diao, S.; Wang, C.; Gong, H.; Liu, T.; Hong, G.; Shi, X.; Dai, H.; Liu, Z. *Adv. Mater.,* 2014, 26, 5646-5652.

[28] Biris, A. S.; Boldor, D.; Palmer, J.; Monroe, W. T.; Mahmood, M.; Dervishi, E.; Xu, Y.; Li, Z.; Galanzha, E. I.; Zharov, V.P. *J Biomed Opt.,* 2009, 14, 021007.

[29] Dhar, S.; Liu, Z.; Thomale, J.; Dai, H.; Lippard, S. J. *J. Am. Chem. Soc.,* 2008, 130,11467-11476.

[30] Zhang, P.; Huang, H.; Huang, J.; Chen, H.; Wang, J.; Qiu, K.; Zhao, D.; Ji, L.; Chao, H. *ACS Appl. Mater. Interfaces*, 2015, 7, 23278-23290.

[31] Chen, Z.; Pierre, D.; Hea, H.; Tan, S.; Pham-Huy, C.; Honga, H.; Huang, J. *Int J Pharm,* 2011, 405, 153-61.

[32] Cheng, J.; Meziani, M. J.; Sun, Y.; Cheng, S. H. *Toxicol Appl Pharmacol.,* 2011, 250, 184-193.

[33] Levi-Polyachenko, N. H.; Merkel, E. J.; Jones, B. T.; Carroll, D. L.; Stewart IV, J. H. *Mol. Pharmaceutics,* 2009, 6, 1092-1099.

[34] Lin, Z.; Liu, Y.; Ma, X.; Hu, S.; Zhang, J.; Wu, Q.; Ye, W.; Zhu, S.; Yang, D.; Qu, D.; Jiang, J. *Scientific Reports,* 2015, 5, 11709.

[35] Ghosh, S.; Dutta, S.; Gomes, E.; Carroll, D.; D'Agostino Jr., R.; Olson, J.; Guthold, M.; Gmeiner, W. H. *ACS Nano*, 2009, 3, 2667-2673.

[36] Samani, M. K.; Khosravian, N.; Chen, G. C. K.; Shakerzadeh, M.; Baillargeat, D.; Tay, B. K. *International Journal of Thermal Sciences,* 2012, 62, 40-43.

[37] Fisher, J. W.; Sarkar, S.; Buchanan, C. F.; Szot, C. S.; Whitney, J.; Hatcher, H. C.; Torti, S. V.; Rylander, C. G.; Rylander, M. N. *Cancer Res.,* 2010, 70, 9855-9864.

[38] Sobhani, Z.; Behnam, M. A.; Emami, F.; Dehghanian, A.; Jamhiri, I. *Int J Nanomedicine,* 2017, 12, 4509-4517.

[39] Merlot, A. M.; Kalinowski, D. S.; Richardson, D. R. *Front Physiol.,* 2014, 5, 299.

[40] Pandey, S.; Talib, A.; Thakur, M. M.; Khan, M. S.; Bhaisare, M. L.; Gedda, G.; Wu, H. *J. Mater. Chem. B,* 2016, 4, 3713-3720.

[41] Mu, X.; Yan, C.; Tian, Q.; Lin, J.; Yang, S. *Int J Nanomedicine,* 2017, 12, 7207-7223.

[42] Moriyama, Y.; Ohta, D.; Hachiya, K.; Mitsui, Y.; Takeda, K. *J Protein Chem.*, 1996, 15, 265-272.

[43] Liu, F.; He, X.; Chen, H.; Zhang, J.; Zhang, H.; Wang, Z. *Nat Commun.,* 2015, 6, 8003.

[44] Zeng, J.; Cheng, M.; Wang, Y.; Wen, L.; Chen, L.; Li, Z.; Wu, Y.; Gao, M.; Chai, Z. *Adv. Healthcare Mater.,* 2016, 5, 772-780.

[45] Jain, P. K.; Lee, K. S.; El-Sayed, I. H.; El-Sayed, M. A. *J. Phys. Chem. B,* 2006, 110, 7238-7248.

[46] Yang, Z.; Song, J.; Dai, Y.; Chen, J.; Wang, F.; Lin, L.; Liu, Y.; Zhang, F.; Yu, G.; Zhou, Z.; Fan, W.; Huang, W.; Fan, Q.; Chen, X. *Theranostics,* 2017, 7, 2177-2185.

[47] Roa, W.; Zhang, X.; Guo, L.; Shaw, A.; Hu, X.; Xiong, Y.; Gulavita, S.; Patel, S.; Sun, X.; Chen, J.; Moore, R.; Xing, J. Z. *Nanotechnology,* 2009, 20, 375101.

[48] Hu, D.; Sheng, Z.; Zhu, M.; Wang, X.; Yan, F.; Liu, C.; Song, L.; Qian, M.; Liu, X.; Zheng, H. *Theranostics*, 2018, 8, 410-422.

[49] Bhattacharya, R.; Mukherjee, P. *Adv. Drug Delivery Rev.,* 2008, 60, 1289-1306.

[50] Hao, E.; Li, S.; Bailey, R. C.; Zou, S.; Schatz, G. C.; Hupp, J. T. *J. Phys. Chem. B*, 2004, 108, 1224-1229.

[51] Kalele, S.; Gosavi, S. W.; Urban, J.; Kulkarni, S. K. *Current Science,* 2006, 91, 1038-1052.

[52] Bardhan, R.; Grady, N. K.; Joseph R. Cole. J. R.; Amit Joshi, A.; Naomi J. Halas, N. J. *ACS Nano,* 2009, 3, 744-752.

[53] Oldenburg, S. J.; Averitt, R. D.; Westcott, S. L.; Halas, N. *J. Chem. Phys. Lett.,* 1998, 288, 243-247.

[54] Jin, Y. *Acc. Chem. Res.*, 2014, 47, 138-148.

[55] Loo, C.; Lowery, A.; Halas, N.; West, J.; Drezek, R. *Nano Lett.,* 2005, 5, 709-711.

[56] Wu, C.; Yu, C.; Chu, M. *Int J Nanomedicine,* 2011, 6, 807-813.

[57] Li, C.; Jamison, A. C.; Rittikulsittichai, S.; Lee, T.; Lee, T. R. *ACS Appl. Mater. Interfaces*, 2014, 6, 19943-19950.

[58] Vongsavat, V.; Vittur, B. M.; Bryan, W. W.; Kim, J.; Lee, T. R. *ACS Appl. Mater. Interfaces,* 2011, 3, 3616-3624.

[59] Noh, M. S.; Lee, S.; Kang, H.; Yang, J.; Lee, H.; Hwang, D.; Lee, J. W.; Jeong, S.; Jang, Y.; Jun, B.; Jeong, D. H.; Kim, S. K.; Lee, Y.; Cho, M. *Biomaterials,* 2015, 45, 81-92.

[60] Hirsch, L. R.; Stafford, R. J.; Bankson, J. A.; Sershen, S. R.; Rivera, B.; Price, R. E.; Hazle, J. D.; Halas, N. J.; West, J. L. *Proc. Natl. Acad. Sci.,* 2003, 100,13549-13554.

[61] Loo, C.; Lin, A.; Hirsch, L.; Lee, M.; Barton, J.; Halas, N.; West, J.; Drezek, R. *Technol. Cancer Res. Treat.,* 2004, 3, 33-40.

[62] Tam, F.; Goodrich, G. P.; Johnson, B. R.; Halas, N. *J. Nano Lett.*, 2007, 7, 496-501.

[63] Bardhan, R.; Chen, W.; Perez-Torres, C.; Bartels, M.; Huschka, R. M.; Zhao, L. L.; Morosan, E.; Pautler, R. G.; Joshi, A.; Halas, N. *J. Adv. Funct. Mater.,* 2009, 19, 3901-3909.

[64] Mackey, M. A.; Ali, M. R. K; Austin, L. A.; Near, R. D.; El-Sayed, M. A. *J. Phys. Chem. B,* 2014, 118, 1319-1326.

[65] Liu, J.; Liang, H.; Li, M.; Luo, Z.; Zhang, J.; Guo, X.; Cai, K. *Biomaterials,* 2018, 157,107-124.

[66] Zhou, M.; Zhang, R.; Huang, M.; Lu, W.; Song, S.; Melancon, M. P.; Tian, M.; Liang, D.; Li, C. *J Am Chem Soc.,* 2010, 132, 15351-15358.

[67] Xia, B.; Wang, B.; Shi, J.; Zhang, Y.; Zhang, Q.; Chen, Z.; Li. J. *Acta Biomaterialia,* 2017, 51,197-208.

[68] Xia, B.; Zhang, Q.; Shi, J.; Li, J.; Chen, Z.; Wang, B. *Colloids Surf B Biointerfaces,* 2018, 164, 291-298.

[69] García-Hevia, L.; Villegas, J. C.; Fernández, F.; Casafont, I.; González,; Rafael Valiente, R.; Fanarraga, M. L. *Adv. Healthcare Mater.,* 2016, 5, 1080-1087.

[70] Iancu, C.; Mocan, L.; Bele, C.; Orza, A. I.; Tabaran, F. A.; Catoi, C.; Stiufiuc, R.; Stir, A.; Matea, C.; Iancu, D.; Agoston-Coldea, L.; Zaharie, F.; Mocan, T. *Int J Nanomedicine,* 2011, 6, 129-141.

[71] Burke, A.; Ding, X.; Singh, R.; Kraft, R. A.; Levi-Polyachenko, N.; Rylander, M. N.; Szot, C.; Buchanan, C.; Whitney, J.; Fisher, J.; Hatcher, H. C.; D'Agostino, Jr., R.; Kock, N. D.; Ajayan, P. M.; Carroll, D. L.; Akman, S.; Torti, F. M.; Torti, S. V. *Proc. Natl. Acad. Sci.,* 2009,106, 12897-12902.

[72] McAtee, C. O.; Barycki, J. J.; Simpson, M. A. *Adv. Cancer Res.,* 2014, 123, 1-34.

[73] Liwen Zhang, L.; Shi Gao, S.; Fan Zhang, F.; Kai Yang, K.; Qingjie Ma, Q.; Lei Zhu, L. *ACS Nano,* 2014, 8, 12250-12258.

[74] Feng, G.; Fang, Y.; Liu, J.; Geng, J.; Ding, D.; Liu, B. *Small,* 2017, 13, 1602807.

[75] Tran, T. H.; Hanh, T. N.; Tran, T. T. P.; Ku, S. K.; Jeong, J.; Choi, H.; Yong, C. S.; Kim, J. O. *Nanomedicine,* 2017, 12, 1511-1523.

[76] Chen, Q.; Wang, C.; Zhan, Z.; He, W.; Cheng, Z.; Li, Y.; Liu, Z. *Biomaterials,* 2014, 35, 8206-14.

[77] Chen, Q.; Liang, C.; Wang, X.; He, J.; Li, Y.; Liu, Z. *Biomaterials,* 2014, 35, 9355-62.

[78] Yu, Q.; Liu, Y.; Cao, C.; Le, F.; Qin, X.; Suna, D.; Liu, J. *Nanoscale,* 2014, 6, 9279-9292.

[79] Song, X.; Chen, Q.; Liu, Z. *Nano Research,* 2015, 8, 340-354.

[80] Mei Chen, M.; Xiaoliang Fang, X.; Tanga, S.; Zheng, N. *Chem. Commun.,* 2012, 48, 8934-8936.

[81] Zhang, Z.; Wang, J.; Xin, N.; Wen, T.; Ji, Y.; Wu, X.; Zhao, Y.; Chen, C. *J. Am. Chem. Soc.,* 2014, 136, 7317–7326.

[82] Yang, J.; Choi, J.; Bang, D.; Kim, E.; Lim, E.; Park, H.; Suh, J.; Lee, K.; Yoo, K.; Kim, E.; Huh, Y.; Haam, S. *Angew. Chem.,* 2011, 123, 461-464.

[83] Zhou, J.; Lu, Z.; Zhu, X.; Wang, X.; Liao, Y.; Ma, Z.; Li, F. *Biomaterials,* 2013, 34, 9584-9592.

[84] Jiang, B.; Zhang, L.; Zhu, Y.; Shen, X.; Ji, S.; Tan, X.; Cheng, L.; Liang, H. *J. Mater. Chem. B,* 2015, 3, 3767-3776.

[85] Geng, J.; Sun, C.; Liu, J.; Liao, L.; Yuan, Y.; Thakor, N.; Wang, J.; Liu, B. *Small,* 2015, 11, 1603-1610.

[86] Zhang, C.; Zhao, X.; Guo, S.; Lin, T.; Guo, H. *Int J Nanomedicine,* 2017, 12, 1827-1840.

[87] Sun, L.; Li, Q.; Zhang, L.; Xu, Z.; Kang, Y.; Xue, P. *ACS Appl. Nano Mater.,* 2018, 1, 325-336.

[88] Li, Z.; Xue-Feng, Y.; Chu, P. K. *J. Mater. Chem. B,* 2018, 6, 1296-1391.

[89] Link, S.; El-Sayed, M. A. *J. Phys. Chem. B,* 1999, 103, 4212-4217.

[90] Li, W.; Zhang, H.; Guo, X.; Wang, Z.; Kong, F.; Luo, L.; Li, Q.; Zhu, C.; Yang, J.; Lou, Y.; Du, Y.; You, J. *ACS Appl. Mater. Interfaces*, 2017, 9, 3354-3367.

[91] Sikdar, D.; Rukhlenko, I.; Cheng, W.; Premaratne, M. *Biomed Opt Express.,* 2013, 4, 15-31.

[92] Yang, J.; Jihye Choi, Bang, D.; Kim, E.; Lim, E.; Park, H.; Suh, J.; Lee, K.; Yoo, K.; Kim, E.; Huh, Y.; Haam, S. *Angew. Chem. Int. Ed.,* 2011, 50, 441-444.

[93] Chenxiao Jiang, C.; Hao Cheng, H.; Ahu Yuan, A.; Xiaolei Tang, X.; Jinhui Wu, J.; Yiqiao Hu, Y *Acta Biomater.,* 2015, 14, 1-9.

[94] Deng, K.; Li, C.; Huang, S.; Xing, B.; Jin, D.; Zeng, Q.; Hou, Z.; Lin, J. *Small,* 2017, 13, 1702299.

In: Nanotechnology
Editor: Paresh Chandra Ray
ISBN: 978-1-53613-889-4

Chapter 5

IRON OXIDE NANOPARTICLES FOR MAGNETIC RESONANCE IMAGING WITH IMPROVED SENSITIVITY

Pohlee Cheah and Yongfeng Zhao*
Department of Chemistry and Biochemistry,
College of Science, Engineering and Technology,
Jackson State University, Jackson, Mississippi, US

ABSTRACT

Due to the unique magnetic properties and excellent bio-compatibility, iron oxide nanoparticles (IONPs) have been widely used in different applications, including separation, hyperthermia therapy, and magnetic resonance imaging (MRI). To utilize the full potential of MRI, recent studies have focused on improving the sensitivity of MRI. This chapter will focus on the recent efforts to develop IONPs based contrast agents to improve the sensitivity of MRI. We will review the development of IONPs to increase transverse relaxivity as T_2-weighted MRI contrast agents. The IONPs with large effective radii, high surface

* Corresponding Author Email: yongfeng.zhao@jsums.edu.

to volume ratio, increased architectural heterogeneity result in T_2 signal enhancement. When the size is ultra-small (<5 nm), IONPs can be used as T_1-weighted contrast agents. Nevertheless, surface ligands on IONPs can affect the chemical exchange between the water molecules and nanoparticles, and change the longitudinal and transverse relaxivities. The ability of IONPs to self-assemble and exhibit T_1-T_2 switching (sharp contrast transition) at different biological environment (i.e., pH) further enhances the IONPs sensitivity. Incorporation of IONPs with radiotracers/photosensitizer enable the use of PET/ SPECT/ fluorescence with MRI to give detailed anatomical and molecular metabolic information. Incorporation of targeting ligands onto IONPs is reported to have specific targeting effect as demonstrated by *In vitro* and *In vivo* studies. The IONPs have shown great potential to be an excellent tool for diagnose and therapy.

Keywords: iron oxide nanoparticles, T_1-T_2 contrast agent, MRI, multi-modal therapy

1. Introduction

Among all nanoparticles, iron oxide nanoparticles (IONPs) gain a huge interest for biomedical application due to their tunable size, unique magnetic property and excellent biocompatibility. Superparamagnetic IONPs emerge as a prominent tool for magnetic resonance imaging (MRI) owning to its colloidal stability *In vivo* and high magnetization in an applied magnetic field. Although gadolinium-based agents are the most commonly used MRI contrast agents in the market, they however carry high toxicity risk in patients particularly with renal dysfunction who is susceptible to nephrogenic systemic fibrosis [1, 2]. On the other hand, IONPs have been used as the more promising and safer alternative because they can be readily metabolized via normal iron blood pool of a biological system.

At the advancement of MRI applications in clinical use, the IONPs based contrast agents are made commercially available with some underwent different stages of clinical trials (Table 1) [3]. For example, Feridex® (Ferumoxides, Feridex® IV, Berlex Laboratories, Endorem®, France) and Resovist® (Ferucarbotran, Resovist®, Bayer Healthcare, Germany) are designed for liver imaging. Ferumoxytol (Feraheme®, AMAG Pharmaceuticals, Cambridge, MA) is used for the treatment of iron deficit anemia. Ferumoxtran-10 (formerly Sinerem®, Guerbet, Roissy, France; and Combidex, AMAG Pharmaceuticals, Cambridge, MA, USA) is primarily designed for lymph node MRI. More detailed update of recent clinical research on IONPs have been reviewed by Wang and Idee [4].

Comparing to different imaging modalities in clinical use, MRI can provide high resolution, no limit to tissue penetration, and no ionic radiation exposure involved. The disadvantage, however, is that the sensitivity of MRI is moderate. Although it is better than that of CT, it is much less sensitive than that of positron emission tomography (PET) or single photon emission computed tomography (SPECT) (Table 2). Hence, the development of IONPs as contrast agents with enhanced sensitivity of MRI will greatly facilitate the further use of MRI in clinical [5]. The ultimate goal is to increase the signal to noise ratio and decrease the usage of amount of contrast agents.

As we are going to discuss in this book chapter, the synthesis and surface modification of iron oxide nanoparticles are crucial for *In vivo* behavior of IONPs. There are already many detailed review papers on synthesis and surface modification of IONPs [3, 6–10] as well as their biomedical applications [11]. This book chapter focuses on the recent efforts to develop IONPs with improved sensitivity for MRI. These efforts include increase of contrast signal, development of T_1-weighted MRI, IONPs that are environmental responsive, and combination with other modalities. In addition to the imaging features, we also summarize the emerging area to develop IONPs for imaging guided therapy. When IONPs are loaded with chemical drug or functionalized with designed molecule, they can be used for imaging guided therapy. These engineered IONPs exhibit novel techniques with the complementary benefits of imaging, delivery and treatment.

Table 1. List of iron oxide contrast agents which are commercialized or at different clinical stages[a]

Compound	Class: Size	Coating agent	$T_{1/2}$	Indication	Development	Trade name
Ferristene (OMP)	SPIO: 3500 nm (composed of crystals below 50 nm)	Sulfonated styrene-divinylbenzene copolymer	Oral	Gastrointestinal	Sale	Abdoscan® (GE-Healthcare)
Ferumoxsil (AMI-121)	SPIO: 300 nm (composed of 10 nm crystals)	Siloxane	Oral	Gastrointestinal		GastroMARK® (Advanced Magnetics); Lumiren® (Guerbet)
Ferrixan (Ferucarbotran, SHU 555A)	SPIO: 60 nm	Dextran	2.4-3.6 h	Liver		Resovist® (Schering); Cliavist™ (Medicadoc)
Ferumoxide (AMI-25, SHU 555A)	SPIO: 120-180 nm; 80-150 nm	Dextran	6 min	Liver		Feridex® (Advanced Magnetics); Endorem™ (Guerbet)
(SHU 555C)	USPIO: ≤ 20 nm	Dextran	6 h	Angiography	Phase 1	Supravist™ (Schering)
Ferumoxtran-10 (AMI-227, BMS-180549)	USPIO: 20-40 nm	Dextran	24-36 h	Lymph node, liver, angiography	Phase 3	Combidex® (Advanced Magnetics); Sinerem® (Guerbet)
Feruglose (PEG-feron, NC 100150)	USPIO: Core size 5-7 nm; total size 20 nm	Carbohydrate-polyethylene glycol coating (PEGylated starch)	6 h	Lymph node, liver, angiography	Preclinical	Clariscan™ (GE-Healthcare)
VSOP-C184)	USPIO: 7 nm	Citrate	0.6-1.3 h	Angiography		

[a]$T_{1/2}$: blood half-life.

Reprinted with permission from ref [3]. Copyright 2009 The Royal Society of Chemistry.

Table 2. Comparison of different imaging modalities

Modality	Advantages	Disadvantages
SPECT	High sensitivity (10^{-14} – fmol) Signal quantification	Low spatial resolution (1-2 mm) Radiation exposure
PET	High sensitivity (10^{-15} – fmol) Signal quantification Use of short half-life radionuclides	Low spatial resolution (1-2 mm) Radiation exposure
CT	High spatial resolution (50-200 μm)	Low sensitivity (10^{-6} – μmol) Radiation exposure
MRI	High spatial resolution (10-100 μm)	Low sensitivity (10^{-9} – nmol) Use of contrast agents
IVUS	Differentiation between layers in the arterial wall High axial resolution (~100 μm) Details about plaque composition	Invasive method by catheter
OCT	Higher resolution than IVUS (~10 μm)	Invasive method by catheter Requires saline flushing of the lumen Limited depth of view
FMT	Versatility of use	Limited depth of view for infrared probes

2. Increase R_2 of Nanomaterials for T_2 Imaging

When IONPs are distributed under a magnetic field, the microscopic field gradients cause diphase and shortening of longitudinal relaxation time (T_1) and transverse relaxation time (T_2 and T_2*) of the nearby nuclei. According to Koenig-Keller model [13], the longitudinal or transverse relaxivity, R_i can be expressed as following:

$$R_i = \frac{1}{T_i^m} = \frac{a}{d_{NP}D}\gamma^2\mu^2 C_{NP} J(\omega, \tau_D)(i = 1,2) \qquad \text{(Equation 1)}$$

where a is a constant, d_{NP} is the diameter of the nanoparticle, D is the diffusion coefficient of water, μ is the magnetic moment of the nanoparticles, γ is the gyromagnetic ratio of the water proton, C_{NP} is the concentration of the nanoparticles, and $J(\omega,\tau_D)$ is the spectral density function. T_1 "light" contrast agent affects the relaxivity through chemical exchange between water and paramagnetic ions (unpaired electrons). While for the T_2 "dark" contrast agent, the relaxivity is highly dependent on the magnetic spin in the nanocrystals structure as well as the unpaired electrons on the surface. In other word, T_1 contrast agent mainly depends on the surface property of the nanoparticles, while T_2 contrast agent is very much affected by size, shape, component, crystal structure, and surface property.

Gao group successfully synthesized size-controllable anisotropic shape/structure IONPs [14, 15] with improved T_2 contrast effect (Figure 1). IONPs with metal-rich surfaces such as $Fe_3O_4\{111\}$ facets (in IO plates, truncated octahedron, tetrahedrons) or $Fe_3O_4\{100\}$ facets (in IO cubes) have 2-4 times higher r_1 values than that of spherical IONPs. In addition to that, IONPs with anisotropic morphology exhibit local magnetic field inhomogeneity resulted in higher r_2 value and enhanced T_2 contrast effects. The r_2 for octapod and cube IONPs are 2 and 5 times larger respectively, compared to that of the spherical IONPs measured at the same conditions due to the larger effective radii and surface-to-volume ratios. Clustered IONPs with increased architectural heterogeneity and reduced symmetry are also found to greatly enhance the magnetic field inhomogeneity, and therefore the r_2 values [16]. In Figure 2, C1 consists of 5 nm-IONPs clustering, C2 consists of 15 nm-IONPs clustering and C3 is made from both 5 nm-IONPs and 15 nm-IONPs (1:1 ratio) clustering. Despite having very close core size (Figure 2a-c) and hydrodynamic size (Figure 2d) for C1, C2 and C3 IONP-clusters, saturated magnetization (M_s) slightly reduces upon clustering as compared to the single component IONPs (Figure 2e). Both the size and dynamic magnetic response are directly proportional to r_2, resulted in 3-fold increase of r_2 upon clustering (Figure 2f). As a result, the clustered IONPs can be used as T_2 contrast agents even at very low iron concentration, greatly enhanced the sensitivities in MR phantom study (Figure 2g).

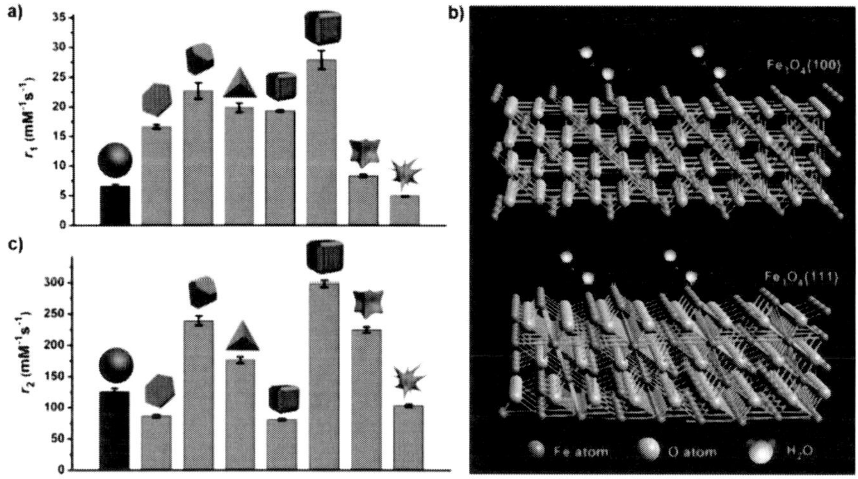

Figure 1. T_1 and T_2 MRI relaxivities of the obtained IONPs, 16 nm sized spherical IONPs (black column, gray sphere) were used for comparison. (a) r_1 values of IO spheres, plates, truncated octahedrons, tetrahedrons, cubes-15 (side length: 15 nm), cubes-21 (side length: 21 nm), concaves, and multibranches structures from left to right. (b) The possible phenomenon of protons chemical exchange on metal-rich surfaces (Fe_3O_4\{100\} and Fe_3O_4\{111\} facets), which indicates that exposed metals on surfaces greatly facilitate chemical exchange of water molecules and accelerate T_1 relaxation. The DMSA molecules on surfaces of IO nanostructures are eliminated for clarity. (c) r_2 values of the selected IO nanostructures, which show that T_2 relaxivity of IO nanostructures is related not only to the saturated magnetization (M_s) value but also to the effective radii (irregular shapes). Reprinted from ref [15].

The r_2 value of IONPs is also found to depend on the different surface coating materials. Mao et al. reported that casein coated IONPs had 2.5 times higher r_2 values compared to amphiphilic polymer coated IONPs of the same core size [17]. They explained that the high water permeability of casein molecules greatly facilitates the diffusion between the water molecules and the particles, resulted in enhanced MRI contrast.

In an earlier study, IONPs conjugated with arginine-glycine-aspartic (RGD) peptide were used as highly sensitive T_2-weighted MRI contrast agents that would bind on $\alpha_v\beta_3$ integrin and enable specific tumor targeting [18]. Recent research by Jasanoff et al. has discovered that the microstructural of peptide sequences binded to IONPs has great influences on the MRI relaxivity properties [19].

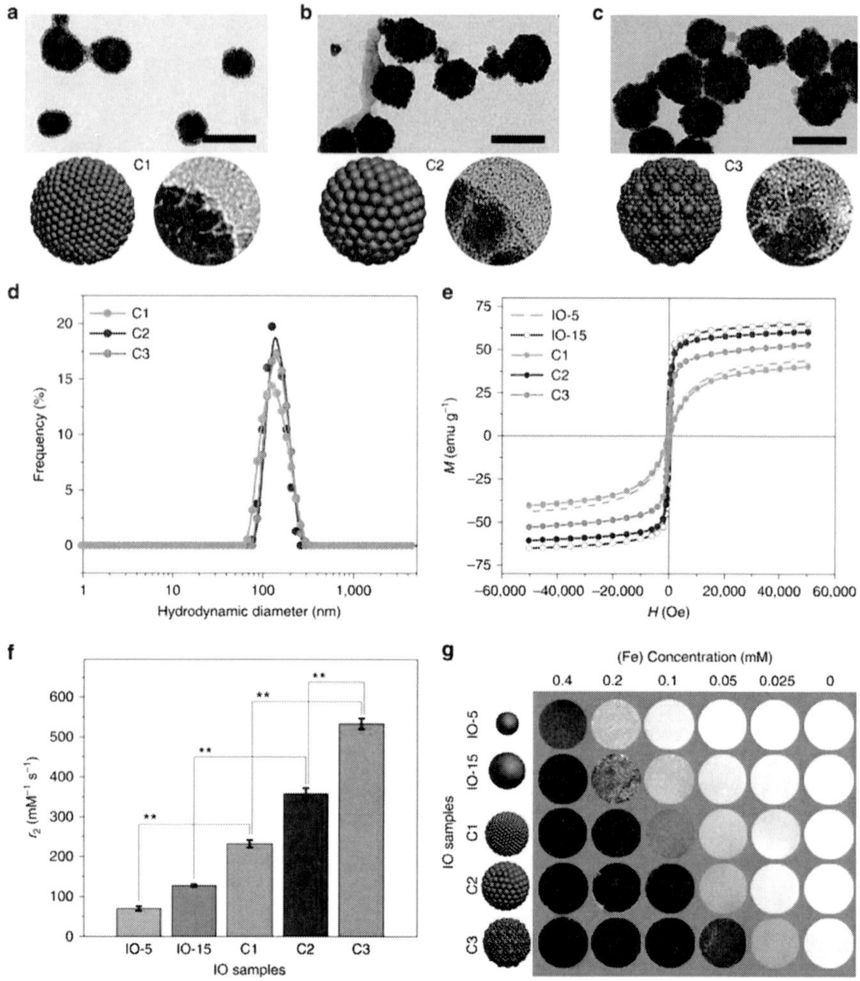

Figure 2. Iron oxide clusters C1–C3. (a–c) Transmission electron microscopy (TEM) and high-resolution TEM images, as well as cartoons of iron oxide (IO) clusters C1–C3, respectively. Scale bars, 200 nm. The inter-particle distances (L) are around 1 nm. (d) Dynamic light scattering (DLS) measurements indicate uniform hydrodynamic diameters of IO clusters C1–C3. (e) Magnetic hysteresis (M–H) curves of single IO-5 and IO-15 NPs, and IO clusters C1–C3 measured at 300K. (f) Columns show r_2 values of IO clusters C1–C3, as well as the single IO-5 and IO-15 NPs. Mean values±s.d.; n = 3 (**P < 0.01). (g) MR phantom of different IO samples show concentration- and relaxivity-dependent contrasts. Darker contrast indicates stronger T_2 relaxation enhancement for each concentration. All MRI studies were conducted on a 7T MRI scanner. Reprinted with permission from ref [16]. Copyright 2017 Springer Nature.

(a)

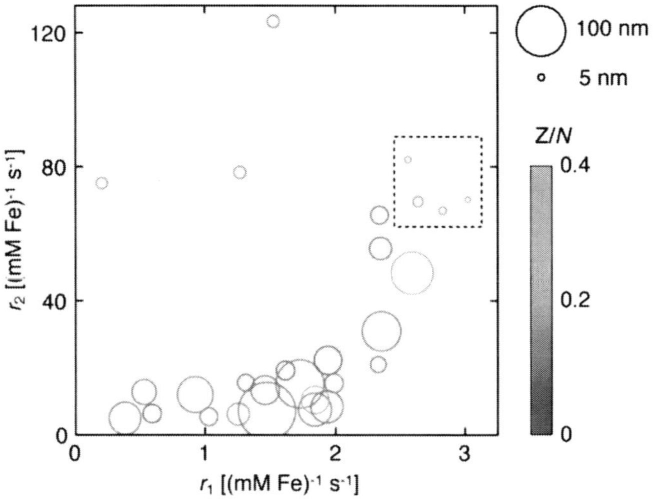

(b)

ligand	r_1		r_2		est. p*I*	*Z/N*
SSSSDDZZ	2.56 ± 0.12		82.3 ± 6.9		3.9	0.25
SSSSDDZZZ	2.63 ± 0.21		69.7 ± 3.6		3.9	0.33
SSSSDDZDZDZ	2.83 ± 0.36		67.1 ± 3.1		3.4	0.27
CDDZ	3.02 ± 0.09		70.3 ± 0.3		3.9	0.25
all others	1.53 ± 0.14		28.5 ± 6.2		4.3 ± 0.2	0.13 ± 0.01

Figure 3. Analysis of peptide–IONP libraries. (a) Characterization of 28 peptide–IONP complexes that passed initial screening for stability from a library of 55 candidates. Each peptide–IONP complex is represented by an open circle centered on coordinates corresponding to its measured relaxivity values (r_1 and r_2) at 7 T. DLS radius measurements from each species are indicated by symbol sizes (key on right), and fractional contents of DOPA residues (*Z/N*) are indicated by the color scale. Individual measurements are reported by peptide sequence in Supporting Information. (b) Properties of optimized contrast agents selected for consistent high r_2 values [dashed box in (a)]. Relaxivity values, estimated p*I*, and fractional DOPA content all differed significantly from the rest of the library. T_1- and T_2-weighted MRI scans illustrating contrast produced by these peptide–IONP complexes (500 μM Fe) are inset at left (*TE/TR* = 10/477) and right (*TE/TR* = 40/3125), respectively. Error margins reflect s.e.m. (n=3). Reprinted with permission from ref [19]. Copyright 2014 American Chemical Society.

The study finds that IONPs with anionic L-3,4-dihydroxyphenylamine (DOPA)-terminated peptide has the greater colloidal stability compared to the previously well-studied citrate ligand. When using much shorter, thiol-functionalized DOPA-terminated peptide the stability is 25% higher than longer peptide or citrate ligand. Among the studied peptide-IONPs complexes, the huge variation of r_2/r_1 ratios ranging from 380 to 4.4 show that the peptide ligand could be designed to optimize the desired interactions of water molecules with IONPs cores (Figure 3).

3. Development of Iron Oxide NPs for T_1-Weighted MRI

Due to the interferences from imaging artifacts or internal bleeding, researchers and clinicians have great interest in developing T_1-weighted (light) over T_2-weighted (dark) MRI contrast agent to improve the imaging procedures. While T_1-weighted contrast agent is highly desirable, clinically available gadolinium-based T_1 agents are found to cause adverse effects in some patients [1, 2]. Therefore, extensive studies of IONPs as the better and safer T_1 agent substitutes have been explored. In tradition, IONPs are used as T_2-weighted (dark) contrast agents but studies have shown the successful use of IONPs as T_1-weighted (light) agents when the size is ultra small (<5 nm).

Taboada et al. synthesized ultra-small IONPs (uIONPs), core sizes around 5 nm with or without surface ligands [20]. While the adsorption of citrate anions on the uIONPs does not alter much r_1 value at 20 MHz, r_2 increases from 28.6 to 35.8 $mM^{-1}s^{-1}$ and r_2/r_1 increased from 1.38 to 2.46. This increase of the ratio r_2/r_1 can be considered as an indicator of the clustering between particles, which could be avoided by optimizing the concentration of citrate. These uIONPs with high magnetization and low r_2/r_1 ratio showed the great potential as T_1 MRI contrast agent. Weller et al. developed an oleic acid and polyethylene glycol (PEG) coated IONPs with core sizes 4 to 6 nm (Figure 4) [21]. When the molecular weight of PEG is greater than 550 g/mol, the resulted IONPs are very stable with increased dispersibility and no aggregation as seen in TEM and DLS. The use of PEG of 1100 g/mol lead to most satisfactory colloidal stability and relaxivity. As the core size of IONPs increase from 4 to 6 nm, both longitudinal and transverse relaxivity increases. The r_1 increases from 7.3 to 13 mM^{-1} s^{-1} while r_2 increases from 17.5 to 42 mM^{-1} s^{-1}. The r_2/r_1 ratio changes from 2.4 to 3.2 show the strong size effect on the relaxivity. The reported r_2/r_1 ratio which is comparable or smaller compared to other conventional contrast agents suggests that PEGylated uIONPs are well suitable for T_1-weighted imaging.

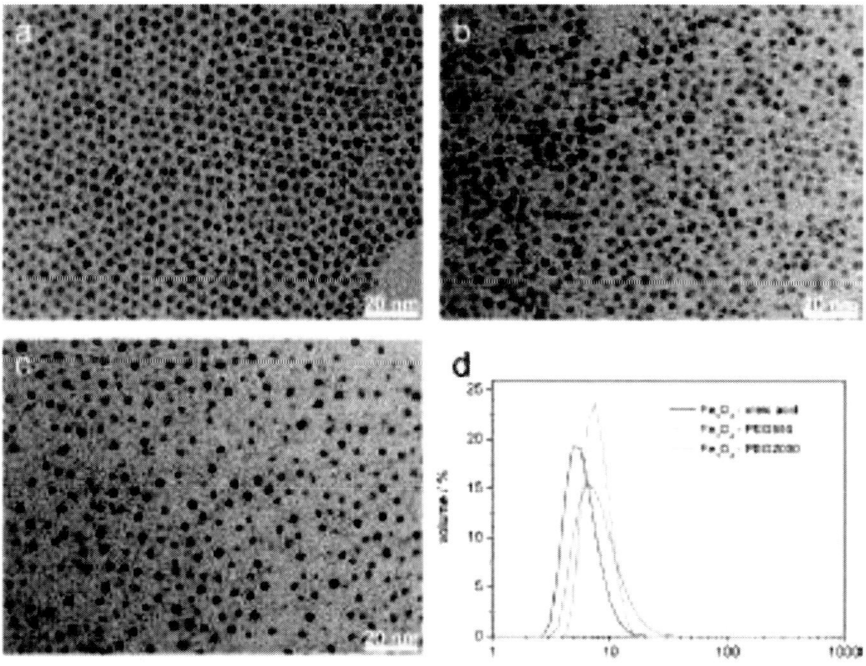

Figure 4. TEM images of the IONPs (4 nm core size) coated with oleic acid (a), PEG 550 (b), and PEG 2000 (c), An increasing distance between the particles with increasing polymer chain length is observable. Together with the DLS results (d) one can conclude that no aggregation occurs during the ligand exchange procedure. Reprinted with permission from ref [21]. Copyright 2009 American Chemical Society.

The surface of uIONPs can be modified with multidentate block copolymer (MDBC) poly(oligo(ethylene oxide) monomethylether methacrylate) (POEOMA) and poly(methacrylic acid) (PMAA) blocks (POEOMA-PMMA) to make a good T_1 contrast agent with core size < 4 nm, r_2/r_1 of 4.74 at 1.4T [22]. Multidendate polymeric ligands consist multiple binding sites to interact with the particles surface and the hydrophilic block copolymer can stabilize uIONPs to give an enhanced colloidal stability and biocompatibility. Even though MDBS-uIONPs has slightly reduced r_1 value, the r_2 value is 2-3 times smaller compared to other dextran-coated uIONPs of core size 5-9 nm. This shows that MDBS-uIONPs have better T_1 contrast enhancement. Another research group also synthesized uIONPs surface functionalized with catechols-grafted POEOMA-b-PMMA ligand. This uIONPs reported to have a core size 3.8 nm, r_1 = 6.8 $mM^{-1}s^{-1}$ and r_2/r_1 = 5.5 at 1.4T, with great T_1 contrast enhancement [23].

Hyeon et al. synthesized uIONPs (< 4 nm) and demonstrated *In vivo* T_1 MR imaging [24]. The 3 nm-sized nanoparticles reported to have a high r_1 relaxivity of 4.8 mM^{-1} s^{-1} and low r_2/r_1 ratio of 6.12. The high r_1 relaxivities of uIONPs can be attributed to the large number of surface Fe^{3+} ions with 5 unpaired valence electrons. In contrary to the fast excretion of gadolinium-based contrast agent (DOTAREM) through the kidney, uIONPs with optimum size ensure sufficient blood circulation time as well as enhancement of the T_1 relaxation (Figure 5). As a result, in-vivo blood pool imaging of the blood vessels as small as 0.2 mm can be achieved. Similar result is reported by Wei et. al. using the zwitterion- (i.e., dopamine sulfonate) coated ultra-small IONPs (ZDS-uIONPs, 3 nm core size) with a maghemite structure [25]. When comparing to other type of IONPs, ZDS-SPIONs have r_1 of 5.2 mM^{-1} s^{-1} and very small r_2/r_1 ratio of 2.0. The r_1 is high enough as a positive T_1 contrast agent for magnetic resonance angiography (MRA) imaging of blood vessels (Figure 6A). Due to very small hydrodynamic size (4.7 nm) and longer blood half-life (~19 min), the ZDS-uIONPs can be cleared by kidney through accumulation at the bladder in reasonable time (Figure 6A-C). They exhibit excellent contrast-to-noise and signal-to-noise ratio between the blood in the vena cava and muscle tissues (Figure 6E-H).

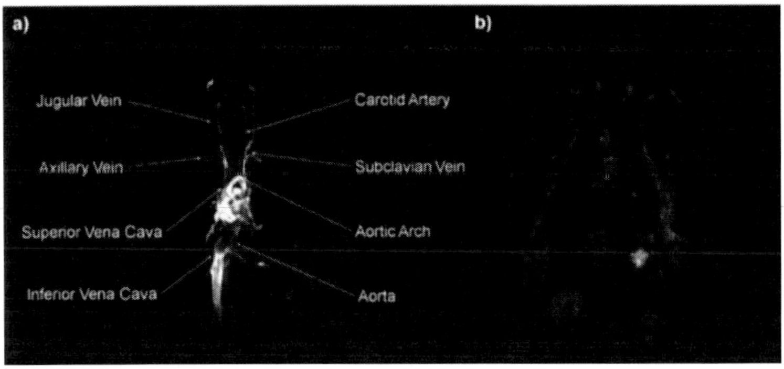

Figure 5. (a) uIONPs- and (b) DOTAREM-enhanced high-resolution blood pool MR images obtained using 3d-FLASH sequence. Reprinted with permission from ref [24]. Copyright 2011 American Chemical Society.

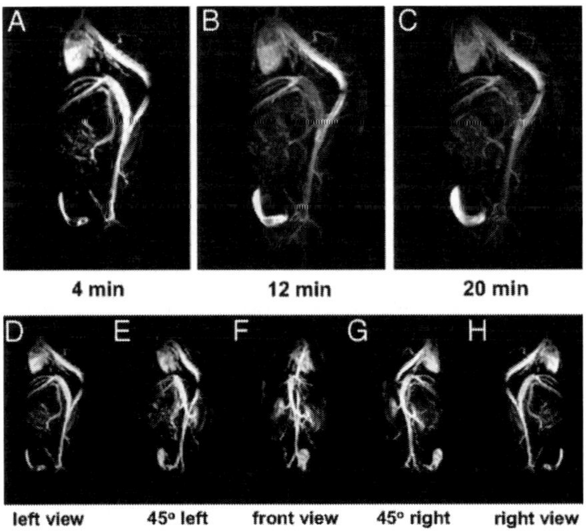

Figure 6. (A–C) T_1-weighted magnetic resonance angiography (MRA) of a mouse injected with ZDS-coated exceedingly small IONPs (ZDS-uIONPs) at 7 T. The time points beneath each image are the time after injection of the ZDS-uIONPs. (D–H) Five different perspectives of the MRA, which are extracted from the 3D scan, at the 4-min mark. A clear positive contrast of the heart and blood vessels are seen, and this positive contrast fades in time while the signal from the bladder increases, consistent with a renal excretion pathway. Reprinted with permission from ref [25]. Copyright 2017 Proceedings of the National Academy of Sciences.

Fan et al. synthesized ultra-small metal ferrite nanoparticles (UMFNPs) by reacting iron-eruciate with cobalt/nickel/manganese-oleate via dynamic simultaneous thermal decomposition and demonstrated the potential use as T_1-MRI contrast agents for blood pool and liver-specific imaging (Figure 7) [26]. Compared to other studied nanoparticles, 2 nm UMFNPs are reported to have the most optimum r_1 relaxivity of 8.43 $mM^{-1}s^{-1}$ and lowest r_2/r_1 of 2.49 at 3T. On top of the size effect of UMFNPs with superior T_1 contrast enhancement for blood vessels MRA imaging, these Mn-containing nanoparticles are hepatocytes-specific for liver imaging. For a better understanding on the clinical translation of uIONPs, large scale uIONPs have been synthesized and the toxicity of these T_1 MRI agents (core size 2 nm, r_2/r_1 of 1.49 at 3T) was studied in non-human primate animal model (Figure 8) [27]. These uIONPs are reported to have excellent imaging performances and biocompatibility and showed the promising evidence of uIONPs as the next generation T_1 contrast agent for clinical trials.

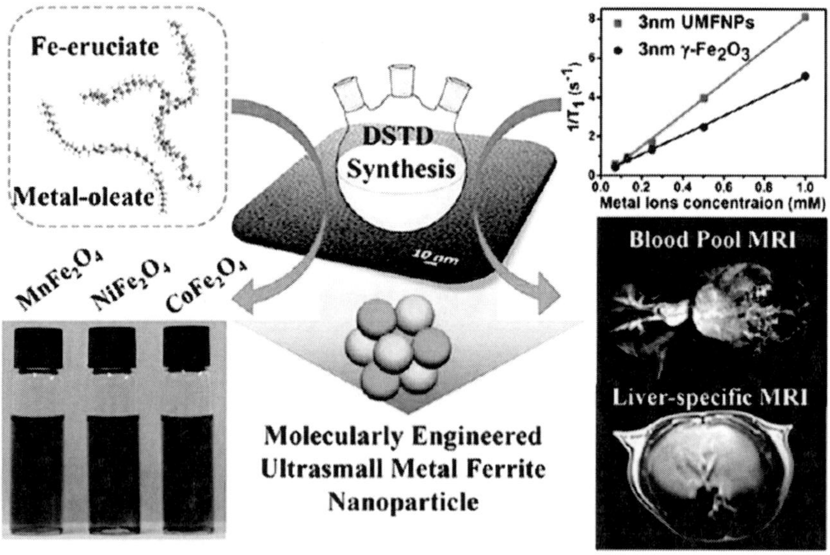

Figure 7. Molecularly engineered ultra-small metal ferrite nanoparticles (UMFNPs) ($MnFe_2O_4$, $NiFe_2O_4$ and $CoFe_2O_4$) synthesized via dynamic simultaneous thermal decomposition (DSTD) for T_1 imaging. Reprinted with permission from ref [26]. Copyright 2017 American Chemical Society.

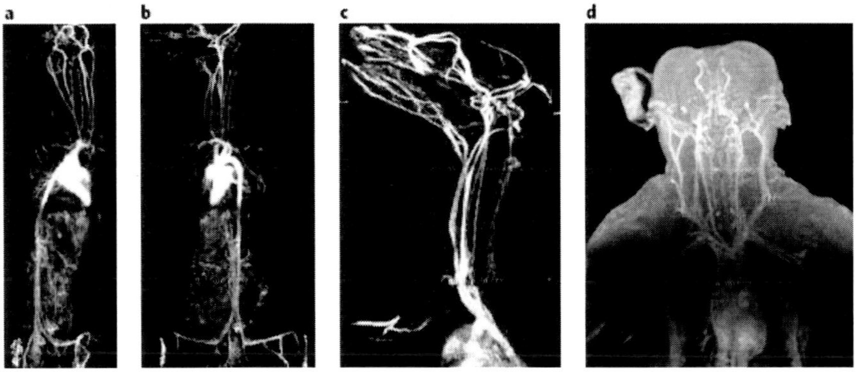

Figure 8. MRA of a beagle dog and a macaque using PEG-uIONPs. a-c, PEG-uIONPs enhanced arterial MRA of the whole body (a,b) and upper body (c) of a beagle dog. d, PEG-uIONPs-enhanced MRA of the upper body of a macaque, showing spatial resolution of the vasculature Reprinted with permission from ref [27]. Copyright 2017 Springer Nature.

4. Tune the MRI Properties Using Self-Assembly

In general, studies showed that self-assembled IONPs (or clustered NPs) exhibit higher MRI contrast effect due to higher effective radii [16]. Mao's group successfully synthesized oligosaccharide-coated uIONPs with sub-5 nm size [28] that could be used as the "light" T_1-MRI agent which readily diffuse in the tumor vasculature. These uIONPs can be used as "dark" T_2-MRI agent when they accumulate and self-assemble/cluster at the acidic tumor sites (Figure 9) [29]. The T_1-T_2 MRI contrast switching is unique and depends on the nanoparticles dispersion form at different tissue environment. They are single-dispersed in the blood/ tumor vasculature but later clustered in the tumor interstitial. This characteristic greatly enhances the extravasation while reducing the intravasation of nanoparticles, and results in excellent permeability and retention effect at the tumor site. The T_1-T_2 MRI contrast switching is an indication that single-dispersed uIONPs (small r_2/r_1 ratio) have self-assembled into clustered IONPs (high r_2/r_1 ratio). For instance, r_2/r_1 ratio for single dispersed uIONPs is 4 and increases to 44 when they clustered. In the MR imaging on liver, strong T_1 contrast by the single dispersed uIONPs at the hepatic vasculatures whereas darken T_2 contrast by the clustered uIONPs at the hepatic sinusoids can be observed at the same time. The authors also find that the T_1-T_2 switching is pH-responsive in which lower pH condition of tumor microenvironments promotes the aggregation/self-assembly of highly dispersed uIONPs into clusters. This can be a contribution from the oligosaccharides coating on the uIONPs that behave like zwitterionics in the presence of both carboxyl and amine functional groups. As the pH decreases, the protonation of carboxyl group leads to aggregation of uIONPs.

Crayton and Tsourkas have synthesized a pH-responsive polymer glycol chitosan (GC) grafted and dextran-stabilized IONPs (GC-IONPs) with core size ~19 nm and a pK_a ~6.5 [30]. GC-IONPs exhibit very significant pH-dependence in T_2^*-weighted MR imaging where r_2 greatly increases as pH decreases. The T_2^*-weighted MR imaging of murine tumor model has shown very significant contrast enhancement in pre- and post-contrast imaging after 24h of GC-IONPs injection. It is believed that the GC-IONPs undergoes highly transition to positive charge upon the encounter of lower pH at the acidic tumor microenvironments, causing GC-IONPs to diffuse and accumulate at the tumor sites.

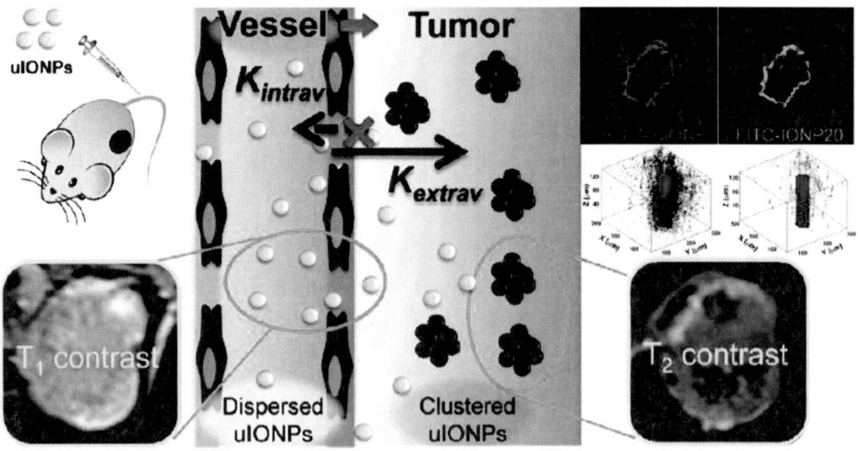

Figure 9. Ultra small iron oxide nanoparticles with T_1-T_2 switchable MRI contrast, enhanced extravasation and reduced intravasation characteristics. Reprinted with permission from ref [29]. Copyright 2017 American Chemical Society.

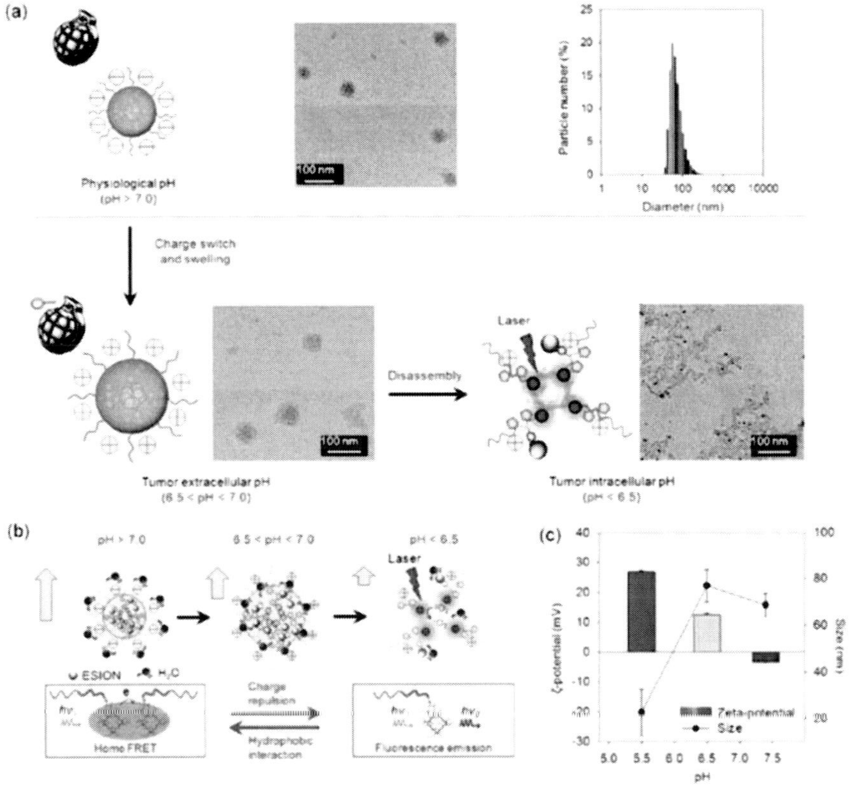

Figure 10. pH-dependent structural transformation and related magnetic/photoactivity change in PMNs. (a) pH-dependent structural transformation behavior in PMNs (inset of (a): TEM images of PMNs at pH 7.4, 6.8, and pH 5.5, DLS measurement of PMN colloidal dispersion at pH 7.4 showing an average diameter of ~70 nm). (b) Schematic representation of pH-dependent structural transformation and related magnetic/ photoactivity change in PMNs. (c) Zeta-potential and DLS size measurement of PMNs (0.1 mg/mL) as a function of pH following a 1 h incubation (n = 3). DLS measurements were conducted using a material refractive index of 2.42, solvent refractive index of 1.33 (water) and a viscosity of 0.8872. Reprinted with permission from ref [31]. Copyright 2014 American Chemical Society.

Hyeon group also introduced the smart ensembles of IONPs assisted by the polymer ligands (i.e., imidazole, pK_a ~6.8) on the surface, which is pH-sensitive to the tumor pH condition of ~5.5 (Figure 10) [31]. These IONPs are termed as pH-sensitive magnetic nanogrenades (PMNs) with core size ~60 nm in which they disassemble into ~3 nm at low pH conditions. The pH dependent structural transformation of PMNs into dispersed uIONPs can be explained by surface charge effect. The ionization of imidazoleas as the pH decreases results in repulsion and complete dissociation. This phenomenon agrees with the relaxivities studies in which the r_2/r_1 ratio decreases along with r_2 as PMNs disassemble at low pH environment. This is strongly supported by the result of T_1 contrast quenching at pH 7.4 but greatly recovered at pH 5.5. The results indicate the excellent properties of PMNs as "light" T_1 MRI contrast agent targeting the acidic tumor sites.

5. Multimodality Imaging

While MR imaging provides great spatial anatomical information, it has rather low sensitivity (microgram range). This prompts the utilization of PET which is known to be highly sensitive (picomolar range) imaging technique. Dual MRI/PET imaging technique enables both the anatomical information and highly sensitive molecular metabolic study at the same time or in real-time.

Lee et al. have utilized the surface amino groups of polyaspartic acid coated on IONPs to couple RGD peptides for integrin $\alpha_v\beta_3$ targeting. At the same time, IONPs are coupled with macrocyclic 1,4,7,10-tetraaza-cyclododecane- N, N', N'', N''', -tetraacetic acid (DOTA) chelators and labeled with ^{64}Cu for PET imaging. ^{64}Cu-DOTA-IONPs-RGD not only provide the anatomical information, they also provide the molecular information of the targeting tumor [18].

Due to the possible dissociation of using the chelators to bind the radionuclides with IONPs, Cai group has developed an effective chelator-free strategy to form *As-IONPs (*= 71, 72, 74, 76) [32]. These nanoparticles are used for sentinel-lymph node mapping which is very critical for cancer metastasis monitoring. The group also synthesized ^{69}Ge-IONPs-PAA/PEG using chelator-free strategy [33]. Results has shown that ^{69}Ge is specifically doped into the core of IONPs and ^{69}Ge adsorption onto surface ligands polyacrylic acid (PAA) is negligible.

While PAA improves the colloidal stability of the IONPs, further functionalization with PEG enhances the *In vitro* serum stability. Accumulation of ^{69}Ge-SPION-PEG in the popliteal lymph nodes can be clearly seen over time, proving the successful uptake of IONPs by the lymphatics system. Herranz et al. have also prepared ^{68}Ga core-doped IONP in which Gallium-68 is incorporated into the core of IONP via microwave synthesis method [34]. This method is proven to be faster and more efficient to synthesize doped IONPs. Further coupling of RGD onto ^{68}Ga-doped IONP demonstrates the successful targeting on $\alpha_v\beta_3$ integrin. Blower et al. have synthesized ^{64}Cu-bis(dithiocarbamatebisphosphonate) ^{64}Cu(dtcbp)$_2$-IONPs for PET/MRI lymphatic imaging [35]. dtcbp is used as an excellent bifunctional chelator which has a dithiocarbamate group to bind with ^{64}Cu and bisphosphonates group to bind to other inorganic materials (i.e., IONPs). Cheon group has developed ^{124}I-SA-MnIONPs for sentinel lymph node study with enhanced sensitivity and resolution [36]. Mn-doped IONPs are first synthesized and their surface is modified with serum albumin (SA) via cross-linking. Radionuclide of ^{124}I is conjugated onto SA-MnIONP through tyrosine residue on SA. The conventional imaging of axillary lymph node is always challenging due to the longer distance away from the injection site and deeper location in the tissues. However, this can now be achieved by overlaying the dual MRI/PET or MRI/SPECT images for detailed study of brachial and axillary lymph node at the same time [37, 38].

Another type of multifunctional nanoprobes is termed as imaging-guided therapy because the IONPs can be used to provide both anatomical information as well as therapy effect. Yin et al. have demonstrated the use of MRI/fluorescence/photothermal therapy (PTT) multifunctional nano-probes for tumor diagnosis and treatment [39]. IONPs are PEGylated with PEG$_{2000}$ and functionalized with folic and photosensitizer chlorin e6 (Ce6). From the MRI and NIR fluorescence study, tumor mass can be located. They then demonstrate the photodynamic therapeutic efficacy of this nanoprobes by treating MGC-803 tumor bearing mice with laser irradiation at 633 nm (50 mW/cm^2) for 30 min. They have observed tumor growth inhibition of ~69% after treating with PEG_{2000}-FA-Ce6-IONPs (Figure 11). Lin et al. have also developed IONPs-polydopamine (PDA) composite that can detect intracellular mRNA (c-myc and TK1) through fluorescence imaging. These IONPs also carry the potential of photoacoustic imaging at 760 nm and effective for PTT treatment (808 nm laser, 2 W/cm^2) on human breast cancer cells MCF-7 (Figure 12) [40]. Photothermal absorption by IONPs-PDA results in temperature increase (from 29.5 to 53.3 °C) as the time of laser irradiation increases in a duration of 10 mins. The rate of heat absorption also increases as the IONPs-PDA concentration increases. From the confocal laser scanning microscopy (CLSM) results, cancer cells are effectively killed after incubation with IONPs-PDA and exposed to the laser. This result is further confirmed by the cytotoxity study in which only ~20% cell viability after incubation of 0.1 mg mL^{-1}IONPs-PDA with laser irradiation compared to 95% cell viability in the control (IONPs-PDA without laser irradiation). Other similar work also reported using IONP-PDA core-shell to target humanized orthotropic lung cancer model (A549), and demonstrated successful PTT treatment [41].

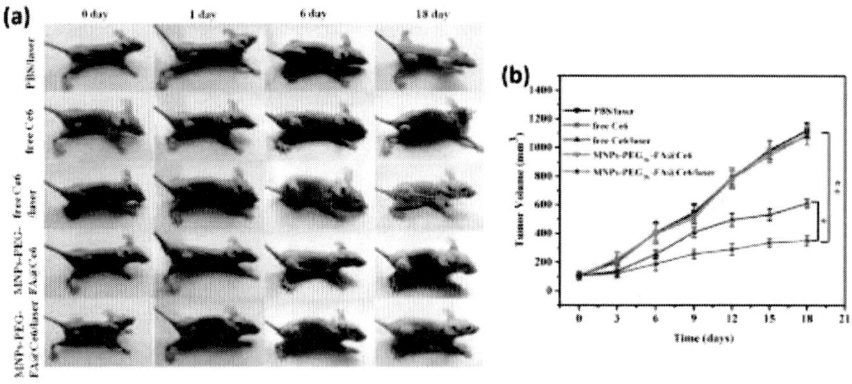

Figure 11. (a) The optical photograph of different groups before and after treatments. The solid tumor sites of tumor-bearing mice respectively treated using PBS, free Ce6 and PEG_{2K}-FA-Ce6-IONPs were irradiated with 633 nm laser (50 mW/cm^2) for 30 min after 4 and 24 h post-injection. (b) MGC-803 tumor growth curves of different groups after treatments. The data are shown as mean ± SD (n = 5).
Reprinted with permission from ref [39]. Copyright 2016 Springer Nature.

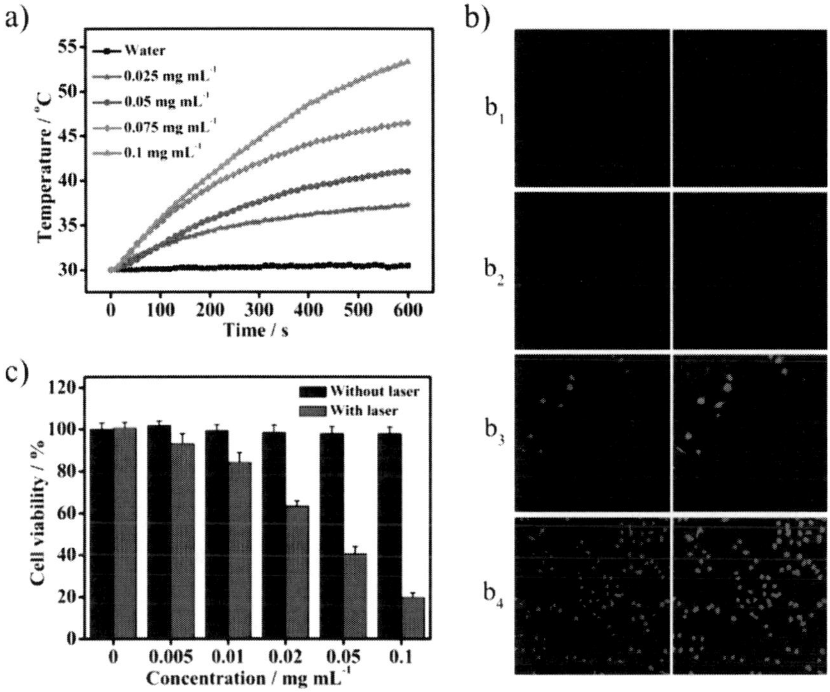

Figure 12. (a) Temperature elevation of different concentrations of IONPs-PDA as a function of irradiation time. (b) CLSM images of differently treated MCF-7 cells stained with PI: (b1) 5 min of laser irradiation only; (b2) IONPs-PDA only; (b3) IONPs-PDA and 2 min of laser irradiation; and (b4) IONPs-PDA and 5 min of laser irradiation. Left panels are PI fluorescence corresponding to dead cells, and right panels are the overlay of PI fluorescence and the bright field image. Scale bars are 50 μm. (c) Cell viability of MCF-7 cells exposed to different concentrations of IONPs-PDA with or without laser irradiation. Reprinted with permission from ref [40]. Copyright 2014 American Chemical Society.

Utilizing the response of epidermal growth factor receptor (EGFR) to overexpressed tyrosine kinase, Wang et al. have developed an anti-EGFR-PEG-IONPs (core size ~9 nm, HD size 40-50 nm) as T_2-MRI contrast agent, targeting for mAb lung cancer H460 tumor cells and greatly enhanced the efficacy of MR-guided focused ultrasound surgery (MRgFUS) ablation treatment [42]. Active targeting of this nanoparticles exhibiting MRI contrast enhancement and ultrasound energy deposition at the target of interest, ensure no damage to the surrounding healthy tissues and organs.

6. Other Strategies to Improve the Sensitivity

As mentioned in the previous sections, T_1-weighted "light" MRI contrast agent gives higher sensitivity due to the interferences/artifacts observed in the T_2-weighted "darken" contrast images. The efforts to improve the imaging sensitivity including optimizations of size and functionalization of the diagnostic tools, selections of specific molecular biomarkers and combination with other optical imaging techniques (multimodality imaging). In addition to the well-established passive tumor delivery of nanoparticles via enhanced permeability retention (EPR) effect where nanoparticles accumulate at the blood vessels that are irregular in shape, dilated and leaky (e.g., lymph nodes, inflammation sites and tumors), specific targeting can improve efficiency and effectiveness of these nanoparticles. The IONPs can be designed for specific targeting to tumor markers, or physiological change of a disease (e.g., protein glycosylation).

To increase the targeting effect, it is desirable to avoid reticulo-endothelial system (RES) uptake of nanoparticles to increase sufficient blood circulation time and increase the sensitivity of the diagnostic nanoprobe. PEGylation is most commonly used to improve the colloidal stability and prevent the uptake by in-situ macrophage cells within the reticuloendothelial system. Sun group has studied the effect of PEGylation for non-specific uptake of IONPs by using a series of PEG diacid molecules of different molecular weight [43]. PEG_{600} –IONPs reported to have 30-50% lower RAW 264.7 macrophage cells uptake than dextran-coated IONPs at the incubation of 0.01 mg Fe mL^{-1}. When the molecular weight is higher, PEG_{3000}, PEG_{6000}, and PEG_{20000}-coated nanoparticles exhibit negligible uptake by the macrophage cells due to the long PEG chain forming dense coating on the surface. Ponti group has studied a different polymer ligand, zwitterions dopamine sulfonate (ZDS) over PEG-like, 2-[2-(2-methoxyethoxy) ethoxy]- acetic acid (MEEA) coating on IONPs [44]. They find that ZDS-coated IONPs has better colloidal stability with no protein adsorptions, very low cellular toxicity, and a specific internalization, indicating the importance of surface chemistry for IONPs delivery. Chester group has also reported prolonged blood circulation time of dextran-coated SPIONs (Ferucarbotran) upon administrations of Fucoidans, a seaweed-derived sulfated polysaccharide which is a known ligand for the scavenger receptor class A (SR A) that is responsible for macrophage uptake [45]. They have observed 4-fold increase of IONPs half-life circulation in the blood (from 37.4 to 150 min) when treated with Fucoidans as compared to treatment with dextran sulfate 500. The result has shown that this ligand has effect in improvement of IONPs blood circulation and thus, the tumor uptake efficiency.

Previously, we introduced some group that demonstrated the studies of IONPs functionalized with RGD peptide that would specifically bind to $\alpha_v\beta_3$ integrin overexpressed cells found in various tumor [18, 34]. Huang et.al. synthesized a specific lung cancer-targeting peptide named H2009.1 (RGDLATLRQL) using PEG_{11} as a linker followed by a cysteine residue at the C-terminus of the peptide to make NH_2-RGDLATLRQL-PEG_{11}-Cys-COOH, or LCP- PEG-Cys ligand [46]. Direct ligand exchange enables the formation of LCP-PEG-Cys-IONPs which can specifically target to the $\alpha_v\beta_6$–positive H2009 cells over $\alpha_v\beta_6$–negative H460 cells. In contrary, there is no uptake of scrambled peptide-IONPs control (SP, with the sequence of DALRLQGTLR-PEG_{11}-Cys-COOH). This result has confirmed that cell targeting can be designed according to the peptide sequence. In addition to that, LCP-PEG-Cys-IONPs also has higher MRI signal enhancement over the SP-IONPs control.

Conclusion and Future Outlook

Extensive studies have been done to develop IONPs for biomedical application. Despite many promising results of using IONPs for theranostics (diagnosis and therapy), the full potential of IONPs is far from being achieved. The primary limitation of MRI is having relatively low sensitivity. In this chapter, we have reviewed recent progress in development of IONPs based contrast agents to improve the sensitivity of MRI. These efforts include: (1) Synthesis of novel IONPs to increase the transversal relaxivity for T_2-weighted MRI; (2) Development of ultra-small IONPs for T_1-weighted MRI; (3) Improving the sensitivity through strategy such as targeted delivery or environmental responding contrast agents, and (4) Combination with other imaging components for multimodality imaging.

Because of capability to be used as T_1-weighted MRI, ultra-small IONPs have emerged as promising contrast agents. The challenges and opportunities for future studies are as following: (1) Robust surface modification methods are needed to gain excellent colloidal stability of IONPs *In vivo* to ensure successful clinical application. (2) Surface materials have reported to affect relaxivity of ultra-small IONPs significantly. A detailed study on surface materials will provide principle to design new T_1-weighted contrast agents. (3) Development of synthetic methods are highly demanded to precisely control the size and structure of ultra-small IONPs to decrease the ratio of r_2/r_1. The goal is to reduce r_2/r_1 to less than 2. Synthesis of ultra-small IONPs with narrow size distributions for T_1-weighted MRI is highly desirable. With effort to address these challenges, it is anticipated that new kinds of IONPs will be developed to improve the sensitivity of MRI, enabling earlier diagnosis of disease and help with designing future treatment plan.

ACKNOWLEDGMENTS

The work is supported by the National Science Foundation under Award No. HRD-1700390.

REFERENCES

[1] P. H. Kuo, E. Kanal, A. K. Abu-Alfa, S. E. Cowper, Gadolinium-based MR Contrast Agents and Nephrogenic Systemic Fibrosis, *Radiology*. 242 (2007) 647–649. doi:10.1148/radiol.2423061640.

[2] J. D. Schlaudecker, C. R. Bernheisel, Gadolinium-associated nephrogenic systemic fibrosis, *Am. Fam. Physician*. 80 (2009) 711–714.

[3] R. Qiao, C. Yang, M. Gao, Superparamagnetic iron oxide nanoparticles: from preparations to *In vivo* MRI applications, *J. Mater. Chem*. 19 (2009) 6274. doi:10.1039/b902394a.

[4] Y. X. J. Wáng, J. M. Idée, A comprehensive literatures update of clinical researches of superparamagnetic resonance iron oxide nanoparticles for magnetic resonance imaging, *Quant. Imaging Med. Surg*. 7 (2017) 88–122. doi:10.21037/qims.2017.02.09.

[5] J. Huang, X. Zhong, L. Wang, L. Yang, H. Mao, Improving the magnetic resonance imaging contrast and detection methods with engineered magnetic nanoparticles, *Theranostics*. 2 (2012) 86–102. doi:10.7150/thno.4006.

[6] C. Boyer, M. R. Whittaker, V. Bulmus, J. Liu, T. P. Davis, The design and utility of polymer-stabilized iron-oxide nanoparticles for nanomedicine applications, *NPG Asia Mater*. 2 (2010) 23–30. doi:10.1038/asiamat.2010.6.

[7] W. Wu, Q. He, C. Jiang, Magnetic iron oxide nanoparticles: Synthesis and surface functionalization strategies, *Nanoscale Res. Lett*. 3 (2008) 397–415. doi:10.1007/s11671-008-9174-9.

[8] S. Laurent, D. Forge, M. Port, a Roch, C. Robic, L. V Elst, R. N. Muller, Magnetic Iron Oxide Nanoparticles: Synthesis, Stabilization, Vectorization, Physicochemical Characterizations, and Biological Applications (vol 108, pg 2064, 2008), *Chem. Rev*. 108 (2008) 2064–2110. doi:Doi 10.1021/Cr900197g.

[9] W. Wu, Z. Wu, T. Yu, C. Jiang, W. S. Kim, Recent progress on magnetic iron oxide nanoparticles: Synthesis, surface functional strategies and biomedical applications, *Sci. Technol. Adv. Mater*. 16 (2015) 23501. doi:10.1088/1468-6996/16/2/023501.

[10] S. K. Yen, P. Padmanabhan, S. T. Selvan, Multifunctional iron oxide nanoparticles for diagnostics, therapy and macromolecule delivery, *Theranostics*. 3 (2013) 986–1003. doi:10.7150/thno.4827.

[11] N. Lee, D. Yoo, D. Ling, M. H. Cho, T. Hyeon, J. Cheon, Iron Oxide Based Nanoparticles for Multimodal Imaging and Magneto-responsive Therapy, *Chem. Rev.* 115 (2015) 10637–10689. doi:10.1021/acs.chemrev.5b00112.

[12] S. M. Kazuma, D. Sultan, Y. Zhao, L. Detering, M. You, H. P. Luehmann, D. S. P. Abdalla, Y. Liu, Recent Advances of Radionuclide-based Molecular Imaging of Atherosclerosis., *Curr. Pharm. Des.* (2015) 5267–5276. doi:10.2174/138161282166615090915104529.

[13] S. H. Koenig, K. E. Kellar, Theory of 1/T1 and 1/T2 NMRD profiles of solutions of magnetic nanoparticles, *Magn. Reson. Med.* 34 (1995) 227–233. doi:10.1002/mrm.1910340214.

[14] Z. Zhao, Z. Zhou, J. Bao, Z. Wang, J. Hu, X. Chi, K. Ni, R. Wang, X. Chen, Z. Chen, J. Gao, Octapod iron oxide nanoparticles as high-performance T2 contrast agents for magnetic resonance imaging, *Nat. Commun.* 4 (2013) 1–7. doi:10.1038/ncomms3266.

[15] Z. Zhou, X. Zhu, D. Wu, Q. Chen, D. Huang, C. Sun, J. Xin, K. Ni, J. Gao, Anisotropic shaped iron oxide nanostructures: Controlled synthesis and proton relaxation shortening effects, *Chem. Mater.* 27 (2015) 3505–3515. doi:10.1021/acs.chemmater.5b00944.

[16] Z. Zhou, R. Tian, Z. Wang, Z. Yang, Y. Liu, G. Liu, R. Wang, J. Gao, J. Song, L. Nie, X. Chen, Artificial local magnetic field inhomogeneity enhances T_2 relaxivity, *Nat. Commun.* 8 (2017) 1–10. doi:10.1038/ncomms15468.

[17] J. Huang, L. Wang, R. Lin, A. Y. Wang, L. Yang, M. Kuang, W. Qian, H. Mao, Casein-Coated Iron Oxide Nanoparticles for High MRI Contrast Enhancement and Efficient Cell Targeting, *ACS Appl. Mater. Interfaces*. 5 (2013) 4632–4639. doi:10.1021/am400713j.

[18] H.-Y. Lee, Z. Li, K. Chen, A. R. Hsu, C. Xu, J. Xie, S. Sun, X. Chen, PET/MRI Dual-Modality Tumor Imaging Using Arginine-Glycine-Aspartic (RGD)-Conjugated Radiolabeled Iron Oxide Nanoparticles, *J. Nucl. Med.* 49 (2008) 1371–1379. doi:10.2967/jnumed.108.051243.

[19] M. Barch, S. Okada, B. B. Bartelle, A. Jasanoff, Screen-based analysis of magnetic nanoparticle libraries formed using peptidic iron oxide ligands, *J. Am. Chem. Soc.* 136 (2014) 12516–12519. doi:10.1021/ja410884e.

[20] M. R. Taboada E, Rodríguez E, Roig A, Oró J, Roch A, Relaxometric and magnetic characterization of ultrasmall iron oxide nanoparticles with high magnetization. Evaluation as potential T1 magnetic resonance imaging contrast agents for molecular imaging., *Langmuir.* 23 (2007) 4583–8.

[21] U. I. Tromsdorf, O. T. Bruns, S. C. Salmen, U. Beisiegel, H. Weller, A Highly Effective, Nontoxic T 1 MR Contrast Agent Based on Ultrasmall PEGylated Iron Oxide Nanoparticles, *Nano Lett.* 9 (2009) 4434–4440. doi:10.1021/nl902715v.

[22] N. Chan, M. Laprise-pelletier, P. Chevallier, A. Bianchi, Multidentate Block-Copolymer-Stabilized Ultrasmall Superparamagnetic Iron Oxide Nanoparticles with Enhanced Colloidal Stability for Magnetic Resonance Imaging, *Bio Macromol.* 15 (2014) 2146–2156.

[23] P. Li, P. Chevallier, P. Ramrup, D. Biswas, D. Vuckovich, M. A. Fortin, J. K. Oh, Mussel-Inspired Multidentate Block Copolymer to Stabilize Ultrasmall Superparamagnetic Fe_3O_4 for Magnetic Resonance Imaging Contrast Enhancement and Excellent Colloidal Stability, *Chem. Mater.* 27 (2015) 7100–7109. doi:10.1021/acs.chemmater.5b03028.

[24] B. H. Kim, N. Lee, H. Kim, K. An, Y. Il Park, Y. Choi, K. Shin, Y. Lee, S. G. Kwon, H. Bin Na, J. G. Park, T. Y. Ahn, Y.-W. Kim, W. K. Moon, S. H. Choi, T. Hyeon, Large-scale synthesis of uniform and extremely small-sized iron oxide nanoparticles for high-resolution T1 magnetic resonance imaging contrast agents., *J. Am. Chem. Soc.* 133 (2011) 12624–12631. doi:10.1021/ja203340u.

[25] H. Wei, O. T. Bruns, M. G. Kaul, E. C. Hansen, M. Barch, A. Wiśniowska, O. Chen, Y. Chen, N. Li, S. Okada, J. M. Cordero, M. Heine, C. T. Farrar, D. M. Montana, G. Adam, H. Ittrich, A. Jasanoff, P. Nielsen, M. G. Bawendi, Exceedingly small iron oxide

nanoparticles as positive MRI contrast agents, *Proc. Natl. Acad. Sci.* 114 (2017) 2325–2330. doi:10.1073/pnas.1620145114.

[26] H. Zhang, L. Li, X. L. Liu, J. Jiao, C. T. Ng, J. B. Yi, Y. E. Luo, B. H. Bay, L. Y. Zhao, M. L. Peng, N. Gu, H. M. Fan, Ultrasmall Ferrite Nanoparticles Synthesized via Dynamic Simultaneous Thermal Decomposition for High-Performance and Multifunctional T1 Magnetic Resonance Imaging Contrast Agent, *ACS Nano*. 11 (2017) 3614–3631. doi:10.1021/acsnano.6b07684.

[27] Y. Lu, Y. J. Xu, G. Zhang, D. Ling, M. Wang, Y. Zhou, Y. D. Wu, T. Wu, M. J. Hackett, B. Hyo Kim, H. Chang, J. Kim, X. T. Hu, L. Dong, N. Lee, F. Li, J. C. He, L. Zhang, H. Q. Wen, B. Yang, S. Hong Choi, T. Hyeon, D. H. Zou, Iron oxide nanoclusters for T 1 magnetic resonance imaging of non-human primates, *Nat. Biomed. Eng.* 1 (2017) 637–643. doi:10.1038/s41551-017-0116-7.

[28] J. Huang, L. Wang, X. Zhong, Y. Li, L. Yang, H. Mao, Facile non-hydrothermal synthesis of oligosaccharides coated sub-5 nm magnetic iron oxide nanoparticles with dual MRI contrast enhancement effect., *J. Mater. Chem. B. Mater. Biol. Med.* 2 (2014) 5311–5482. doi:10.1039/C4TB00811A.

[29] L. Wang, J. Huang, H. Chen, H. Wu, Y. Xu, Y. Li, H. Yi, Y. A. Wang, L. Yang, H. Mao, Exerting Enhanced Permeability and Retention Effect Driven Delivery by Ultrafine Iron Oxide Nanoparticles with T_1-T_2 Switchable Magnetic Resonance Imaging Contrast, *ACS Nano.* 11 (2017) 4582–4592. doi:10.1021/acs nano.7b00038.

[30] S. H. Crayton, A. Tsourkas, PH-titratable superparamagnetic iron oxide for improved nanoparticle accumulation in acidic tumor microenvironments, *ACS Nano*. 5 (2011) 9592–9601. doi:10.1021/nn202863x.

[31] D. Ling, W. Park, S. J. Park, Y. Lu, K. S. Kim, M. J. Hackett, B. H. Kim, H. Yim, Y. S. Jeon, K. Na, T. Hyeon, Multifunctional tumor pH-sensitive self-assembled nanoparticles for bimodal imaging and treatment of resistant heterogeneous tumors, *J. Am. Chem. Soc*. 136 (2014) 5647–5655. doi:10.1021/ja4108287.

[32] F. Chen, P. A. Ellison, C. M. Lewis, H. Hong, Y. Zhang, S. Shi, R. Hernandez, M. E. Meyerand, T. E. Barnhart, W. Cai, Chelator-free synthesis of a dual-modality PET/MRI agent, *Angew. Chemie - Int. Ed.* 52 (2013) 13319–13323. doi:10.1002/anie.201306306.

[33] R. Chakravarty, H. F. Valdovinos, F. Chen, C. M. Lewis, P. A. Ellison, H. Luo, M. E. Meyerand, R. J. Nickles, W. Cai, Intrinsically germanium-69-labeled iron oxide nanoparticles: Synthesis and in-vivo dual-modality PET/MR zimaging, *Adv. Mater*. 26 (2014) 5119–5123. doi:10.1002/adma.201401372.

[34] J. Pellico, J. Ruiz-Cabello, M. Saiz-Alía, G. del Rosario, S. Caja, M. Montoya, L. Fernández de Manuel, M. P. Morales, L. Gutiérrez, B. Galiana, J. A. Enríquez, F. Herranz, Fast synthesis and bioconjugation of ^{68}Ga core-doped extremely small iron oxide nanoparticles for PET/MR imaging, *Contrast Media Mol. Imaging.* 11 (2016) 203–210. doi:10.1002/cmmi.1681.

[35] R. Torres Martin de Rosales, R. Tavaré, R. L. Paul, M. Jauregui-Osoro, A. Protti, A. Glaria, G. Varma, I. Szanda, P. J. Blower, Synthesis of ^{64}Cu^{II}-Bis(dithiocarbamatebisphosphonate) and Its Conjugation with Superparamagnetic Iron Oxide Nanoparticles: *In vivo* Evaluation as Dual-Modality PET-MRI Agent, *Angew. Chemie Int. Ed.* 50 (2011) 5509–5513. doi:10.1002/anie.201007894.

[36] J. Choi, J. C. Park, H. Nah, S. Woo, J. Oh, K. M. Kim, G. J. Cheon, Y. Chang, J. Yoo, J. Cheon, A Hybrid Nanoparticle Probe for Dual-Modality Positron Emission Tomography and Magnetic Resonance Imaging, *Angew. Chemie Int. Ed.* 47 (2008) 6259–6262. doi:10.1002/anie.200801369.

[37] B. Y. Yang, S.-H. Moon, S. R. Seelam, M. J. Jeon, Y. S. Lee, D. S. Lee, J.-K. Chung, Y. Il Kim, J. M. Jeong, Development of a multimodal imaging probe by encapsulating iron oxide nanoparticles with functionalized amphiphiles for lymph node imaging, *Nanomedicine.* 10 (2015) 1899–1910. doi:10.2217/nnm.15.41.

[38] S. Liu, B. Jia, R. Qiao, et al., A Novel Type of Dual-Modality Molecular Probe for MR and Nuclear Imaging of Tumor:

Preparation, Characterization and *In vivo* Application, *Mol. Pharm.* 6 (2013) 1074–1082. doi:10.1039/b902394a.(18).

[39] T. Yin, P. Huang, G. Gao, J. G. Shapter, Y. Shen, R. Sun, C. Yue, C. Zhang, Y. Liu, S. Zhou, D. Cui, Superparamagnetic Fe_3O_4-PEG2K-FA@Ce6 Nanoprobes for *In vivo* Dual-mode Imaging and Targeted Photodynamic *Therapy, Sci. Rep.* 6 (2016) 1–12. doi:10.1038/srep36187.

[40] L. S. Lin, Z. X. Cong, J. B. Cao, K. M. Ke, Q. L. Peng, J. Gao, H. H. Yang, G. Liu, X. Chen, Multifunctional Fe_3O_4 @ Polydopamine Core À Shell Nanocomposites for Intracellular mRNA Detection and, *ACS Nano*. 8 (2014) 3876–3883. doi:10.1021/nn500722y.

[41] R. Zheng, S. Wang, Y. Tian, X. Jiang, D. Fu, S. Shen, W. Yang, Polydopamine-Coated Magnetic Composite Particles with an Enhanced Photothermal Effect, *ACS Appl. Mater. Interfaces*. 7 (2015) 15876–15884. doi:10.1021/acsami.5b03201.

[42] Z. Wang, R. Qiao, N. Tang, Z. Lu, H. Wang, Z. Zhang, X. Xue, Z. Huang, S. Zhang, G. Zhang, Y. Li, Active targeting theranostic iron oxide nanoparticles for MRI and magnetic resonance-guided focused ultrasound ablation of lung cancer, *Biomaterials.* 127 (2017) 25–35. doi:10.1016/j.biomaterials.2017.02.037.

[43] J. Xie, C. Xu, N. Kohler, Y. Hou, S. Sun, Controlled PEGylation of monodisperse Fe_3O_4 nanoparticles for reduced non-specific uptake by macrophage cells, *Adv. Mater*. 19 (2007) 3163–3166. doi:10.1002/adma.200701975.

[44] S. Mondini, M. Leonzino, C. Drago, A. M. Ferretti, S. Usseglio, D. Maggioni, P. Tornese, B. Chini, A. Ponti, Zwitterion-Coated Iron Oxide Nanoparticles: Surface Chemistry and Intracellular Uptake by Hepatocarcinoma (HepG2) Cells, *Langmuir.* 31 (2015) 7381–7390. doi:10.1021/acs.langmuir.5b01496.

[45] M. R. A. Abdollah, T. J. Carter, C. Jones, T. L. Kalber, V. Rajkumar, B. Tolner, C. Gruettner, M. Zaw-Thin, J. Baguña Torres, M. Ellis, M. Robson, R. B. Pedley, P. Mulholland, R. T. M. de Rosales, K. A. Chester, Fucoidan Prolongs the Circulation Time of Dextran-Coated

Iron Oxide Nanoparticles, *ACS Nano*. (2018) acsnano.7b06734. doi:10.1021/acsnano.7b06734.

[46] G. Huang, C. Zhang, S. Li, C. Khemtong, S. G. Yang, R. Tian, J. D. Minna, K. C. Brown, J. Gao, A novel strategy for surface modification of superparamagnetic iron oxide nanoparticles for lung cancer imaging, *J. Mater. Chem.* 19 (2009) 6367. doi:10.1039/b902358e.

In: Nanotechnology
Editor: Paresh Chandra Ray
ISBN: 978-1-53613-889-4

Chapter 6

Multi-Color Carbon Dots for Bioimaging

Salma Begum, Avijit Pramanik, Aruna Vangara and Paresh Chandra Ray*

Department of Chemistry, Physics and Atmospheric Science,
Jackson State University, Jackson, Mississippi, US

Abstract

Multi-color carbon dots (CDs) are the new type of fluorescent carbon materials showing promising potentials in bioimaging and drug delivery because of their low toxicity, high biocompatibility, favorable photostability, tunable fluorescent properties and ultra compact size. This chapter highlights the critical review on the recent development of multicolor carbon dots based nanotechnology, which has been explored for bioimaging of virus and bacteria, cancer cells and brain cells.

Keywords: carbon dots (CDs), multicolor, fluorescent, bioimaging

---

* Corresponding Author Email: paresh.c.ray@jsums.edu.

INTRODUCTION

Carbon is one of the most important elements and major building block of organic compounds. According to the Web of Science database, carbon is the most extensively studied element of the periodic table and is an essential component for all living organisms on this Earth. In addition, carbon plays a key role in the energy and chemical industries. Therefore, it is increasingly being used in diverse areas of chemical research. Carbon quantum dots or carbon dots (CDs) have been recently developed as new type of fluorescent carbon materials from the nano carbon family [1, 2]. These carbon dots are highly fluorescent that have an average size in the range of 1-10 nm. They generally possess a sp^2 conjugated core and contain suitable oxygen content in the forms of multiple oxygen containing species represented by carboxyl, hydroxyl and aldehyde groups. One of the most characteristic properties of carbon dots is the size dependence of the photo physical properties, which allows property control simply by nanoparticle growth. Their emission color can be tuned by varying the excitation wavelength, and are widely used for multicolor patterning and sensing probes (Figure 1). For instance, they can act as fluorescent sensing probes for metal ion detection by complexation through their strong binding and chelating affinity due to abundant surface carboxyl, amine and hydroxyl groups. Surface modification is a usual approach to expand the analytical application and target detection of carbon dots (CDs). These carbon dots have come forth as potential competitors for inorganic quantum dots (QDs) and other toxic, heavy metal based materials because of their unique characteristics such as elemental abundance, low toxicity, eco friendly and biocompatibility [3-5].

Carbon dots were accidentally discovered in 2004 by Xu et al. as a byproduct of crude-arc discharge soot purification [6]. Later in 2006, Sun et al. synthesized them by laser ablation method [7]. Since then carbon dots have been gained a considerable attention as new class of highly luminescent quantum dots with their characteristics advantages. Apart from their strong and tunable fluorescence features carbon dots are superior in terms of their solubility, functionalizability and photostability.

The favorable fluorescence properties of carbon dots also exhibited a great potential for the applications in environmental and biological sensing and imaging. Some applications of carbon dots, such as drug delivery, gene delivery, sensors, antibacterial activity, biosensing and bioimaging are briefly discussed in the subsequent sections (Figure 2) [8-11].

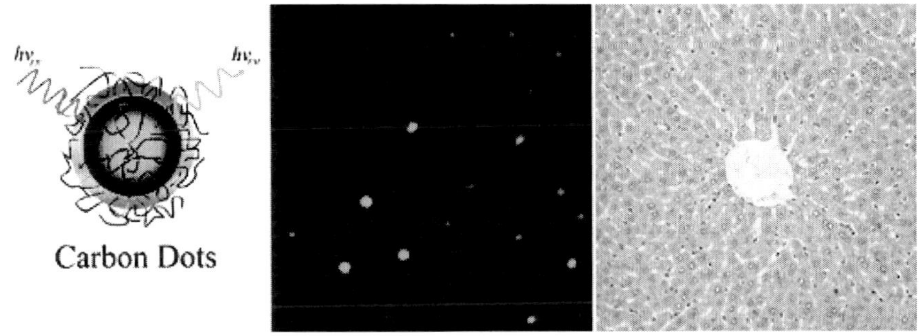

(a) CDs (b) Fluorescent CDs (c) Cell imaging with CDs

Figure 1. Carbon dots (Reproduced with permission from reference [3]. Copyright 2009 American Chemical Society).

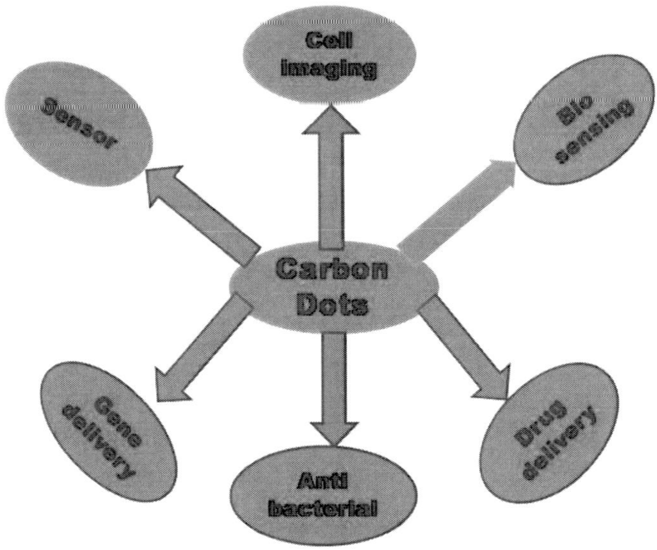

Figure 2. Different applications of carbon dots.

Even though some of the synthetic methods are capable of fabricating a high brightness of carbon dots through surface passivation, the broad emission band of the synthesized carbon dots is still observed because of the lack of uniform size. Therefore, it remains a challenge to create a simple, easy and cheap green strategy for preparation of fluorescent nonmaterial with high fluorescence brightness and narrow emission band. Distinguished by "bottom-up" and "top-down" cutting routes, there are numerous methods that have been developed to synthesize such carbon dots. The fluorescence behavior of carbon dots is usually believed to be originating due to the radiative recombination of excitation in core and surface electronic states, surface functional groups and surface energy traps and the quantum size confinement effect [12-14].

Synthetic methods for the development of carbon dots are classified into three major categories: (1) The cutting of coal, multiwall carbon targets, single wall carbon targets and graphite through carbonization and acidic oxidation etching, laser ablation, electrochemical shocking, arc discharge and electrochemical exfoliation, respectively [6-8]. (2) Thermal, ultrasonic and microwave treatment of molecular precursors, such as ammonium citrate, carbohydrates, glycerol, dopamine and diethylene glycol. Moreover, large quantities of highly fluorescent dots can be obtained in just two minutes using a domestic microwave oven [8-10]. (3) Green synthesis of carbon dots from biological materials like candle soot, green tea, orange peel waste, chicken egg and overcooked barbecue foods [15-19].

SYNTHETIC METHODS

Physical Methods

Laser Ablation Method

Laser ablation synthesis of carbon dots is one of the convenient synthetic methods reported by Sun et al. utilizing a Q-switched Nd: YAG laser on a carbon target [7]. They used graphite powder and cement in the presence of water vapor and argon gas. The obtained carbon nanoparticles exhibited a size spread without displaying any detectable fluorescence. However, after refluxing with HNO_3 for 12 hours using diammine terminated oligomeric polyethylene glycol molecules (PEG), the obtained carbon dots exhibited a bright fluorescence in both solution and solid state.

A single step route to create fluorescent carbon dots by laser irradiation of graphite powder in organic solvents was synthesized by Hu et al. [16]. Graphite powder dispersed in different solvents, such as diamminehydrate, diethanolamine and PEG was irradiated with a Nd: YAG pulsed laser, with ultrasonification of the mixture. The resulting homogeneous black suspension was centrifuged and the colorful supernatant containing carbon dots was separated for further analysis. This group proposed that creating high temperature and pressure by the pulsed laser transferred carbon dots into a plasma state, the formation of carbon dots formed after cooling the mixture [16-20].

Graphite Powder + PEG200N →(LASER) = Carbon Dots

Arc Discharge Method

Xu.et al. demonstrated the separation of carbon dots from single walled carbon nanotubes that were prepared from arc discharge soot, via a preparative electrophoretic method [6]. Arc discharged soot was oxidized with HNO_3 and extracted with basic water. Agarose gel and glass bead electrophoresis were also used to separate carbon dots, which fluoresced different colors under UV light excitation [6, 21].

Plasma Treatment

Li et al. reported a plasma induced synthesis method for blue emitting carbon dots [22]. An acrylamide was placed in a glass dish and treated with air plasma (CTP-2000K) for 10 minutes at radio frequency power of 150 W. The resultant dark brown powder was ultrasonicated in ethanol, centrifuged and filtered to separate the brown supernatant that contained fluorescent carbon dots [22, 23].

Chemical Methods

Electrochemical Synthesis

An electrochemical oxidation is a process involving the use of an oxidative agent which breaks down the carbon precursor into smaller nanoparticles. The carbon precursor gets oxidized and various functional groups are formed, resulting into enhanced solubility and photo-luminescence of the carbon dots. Liu et al. first showed the synthesis of carbon dots by oxidative treatment of soot [24]. It was collected from a burning candle, using HNO_3 as an oxidative agent. The collected soot was hydrophobic and composed of 92% carbon. The collected soot was refluxed with HNO_3 and the reaction mixture was centrifuged. The light brown supernatant thus formed exhibited yellow fluorescence at 312 nm wavelength. They also used other oxidative agents including H_2O_2 and AcOH that showed blue fluorescence. The fluorescent carbon dots were purified using polycrylamide gel electrophoresis (PAGE) that showed multicolor carbon dots such as violet-blue, green-yellow and orange-red with different ranges (which could be either charge of diameter of CDs [24, 25].

Combustion/Hydrothermal/Solvothermal

Peng et al. synthesized carbon dots using carbohydrates via hydrothermal synthesis [26]. The carbohydrates were dehydrated with H_2SO_4 and then treated with HNO_3 which broke down the carbonaceous material into smaller carbon nanoparticles with weak fluorescence. The nanoparicles were then surface-passivated with 7, 10-trioxa-1,13-tridecanediamine which exhibited bright fluorescence and dramatic change in quantum yield. Liu et al. synthesized carbon dots from candle soot by hydrothermal reaction [27]. Collected candle soot was sonicated in NaOH solution and heated. After cooling the reaction mixture, the supernatant was collected, and neutralized with HCl, followed by dialysis. The resulting product contained carbon dots with an average size of 3 nm. Zeta potential measurements confirmed that the carbon dots possess negatively charged surface. Jiang et al. reported the use of solvothermal method for synthesizing multicolor carbon dots from ortho, meta and para-phenylenediamine as precursors, which exhibited green, blue and red photoluminescence respectively [28]. These carbon dots were purified by silica gel chromatography. TEM and AFM measurements revealed mono dispersed carbon dots with an average size of carbon 8.2, 6.0, and 10.0 nm, respectively, with particle height around 2-3 nm.

Wang et al. reported a synthetic method that is called hot injection synthesis method [29]. During this process, citric acid as a carbon precursor was carbonized to form oil-soluble carbon dots, by using octadecene as non-coordinating solvent and 1-hexadecylamine as passivating agent at 300°C under inert atmosphere. Changing the solvent from octadecene to glycerin, they synthesized water-soluble carbon dots in the presence of PEG^{1500N} [29, 30].

Microwave/Induction-Heater Assisted Synthesis

Zhu et al. reported an economical and easy microwave pyrolysis method to synthesize carbon dots within few minutes [17]. Various amounts of PEG-200 with glucose and fructose were dissolved in water and the mixture was heated in a microwave oven at 500 W for 2-10 minutes. The change of color indicated the formation of carbon dots. These carbon dots showed amorphous nature that contained C = O, C = C, C-O-C, C-H and O-H bonds. Wang et al. also reported a microwave synthesis procedure using carbohydrates as carbon precursors, where small amount of inorganic ions were added instead of passivating agents [31]. This represents a simple green synthesis route without the use of any sophisticated instruments.

CHARACTERIZATION OF MULTI COLOR CARBON DOTS

The characterization of carbon dots are performed by standard spectroscopic techniques UV-Vis, fluorescence and FTIR spectroscopy as well as by DLS, XPS, AFM, TEM and HRTEM. The Fourier transform infrared spectroscopy (FTIR) is a useful technique to identify several functional groups, such as $-NH_2$ and -COOH, on the surface of carbon dots (Figure 3). The major carbon source, such as cysteine can also be recognized by the symmetric bending of primary amines and stretching vibrations of thiol groups. These functional groups revealed excellent properties and transformed the carbon dots more dispersible in water, providing good platforms for their conjugation with biomolecules. Transmission electron microscopy (TEM) images and dynamic light scattering (DLS) demonstrated that carbon dots are generally uniformly dispersed with particle diameters of 2.5 to 10 nm (Figure 4). HRTEM images reveal the average diameter of CDs 2.7 nm with relative standard deviation (RSD) of particle size of 9% (n = 100) (Figure 5). An AFM image reveals that topographic heights of the carbon dots are primarily distributed in the range of 2.2 to 3.3 nm. These results reflect that the carbon dots are monodispersed and uniform in size, which is consistent with the result obtained from a TEM image of the carbon dots. To further confirm the near monodispersity of the carbon dots, bulk samples can be analyzed using dynamic light scattering (DLS) [8-13, 32-35].

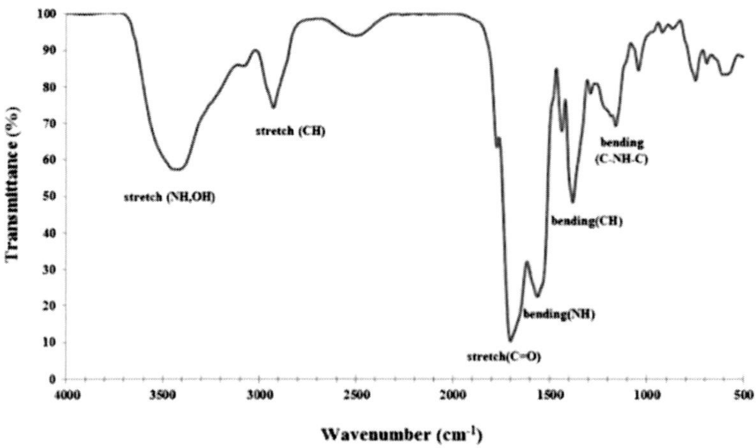

Figure 3. A typical FTIR spectrum of carbon dots. (Reproduced with permission from reference [34]. Copyright 2017 Springer Nature).

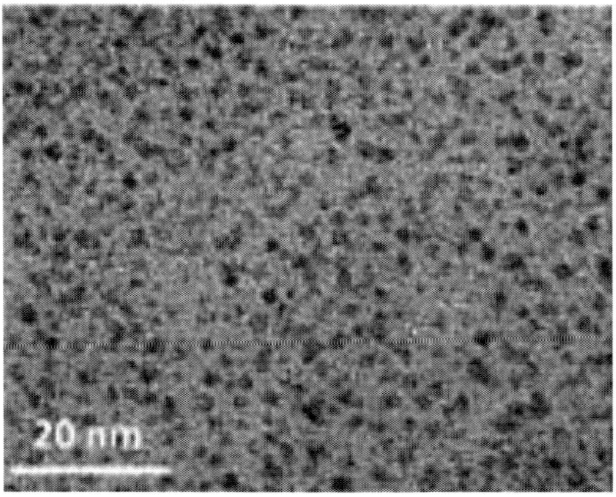

Figure 4. Transmission Electrorn Microscopy (TEM) image of CDs. (Reproduced with permission from reference [35]. Copyright 2017 American Chemical Society).

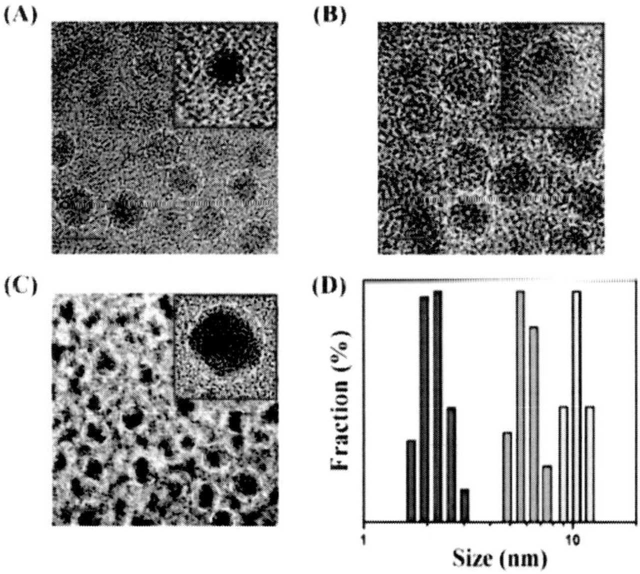

Figure 5. HRTEM images of multicolor CDs synthesized under precursor solutions containing different concentrations of NaOH. (A) 0.4 M, (B) 0.2 M, (C) 0.1 M (insert scales bar are all 2 nm). (D) The corresponding CDs size distribution histogram obtained through Dynamic light scattering (DLS) measurements. The blue, green and yellow columns represent the hydrodynamic size distribution of Blue CDs (1.7-3.5 nm), Green CDs (4.8-7.5 nm), and Yellow CDs (8.7-11.7 nm) respectively. (Reproduced with permission from reference [13]. Copyright 2017 American Chemical Society).

OPTICAL PROPERTIES OF CARBON DOTS

Due to the small size of synthesized carbon dots, UV-Vis and fluorescence spectrum can be used to explain the unique optical properties of carbon dots' multicolor characteristics. The absorption of carbon dots is mostly observed in UV range, which is investigated to the core and surface state transitions, n-π^* and π-π^* C = C transitions, other functional groups and quantum size effects. All of the carbon dots materials consist of sp^2 hybridized carbon atoms possessing π-electrons that can undergo optical transitions in the UV-Vis to NIR region. Moreover, reducing the dimension and symmetry of a bulk material typically changes its optical properties [25-35].

Absorbance

The absorption of carbon dots varies with the different synthetic procedures. The carbon dots synthesized from different techniques may have variable sizes, providing different optical properties. During a reaction process, a precursor's concentration has a close relationship with resulting emitted color of carbon dots. Wang et al. reported one step hydrothermal refluxing process to produce multicolor carbon dots [20]. Cysteine and galactose were chosen as source materials and the reaction was performed with different concentration of NaOH (0.4 M, 0.2 M and 0.1 M) to produce blue, green and yellow carbon dots. Higher NaOH concentration leads to a longer wavelength fluorescence emission (Figure 6). However, decrease of an emission intensity can be observed when different excitation wavelengths were engaged. However, it remains unaffected by the variation of emission wavelength shifts. This observation reveals that the multi-color carbon dots do not have excitation dependent property, which can be attributed to carbon dots surface uniformity [13-20].

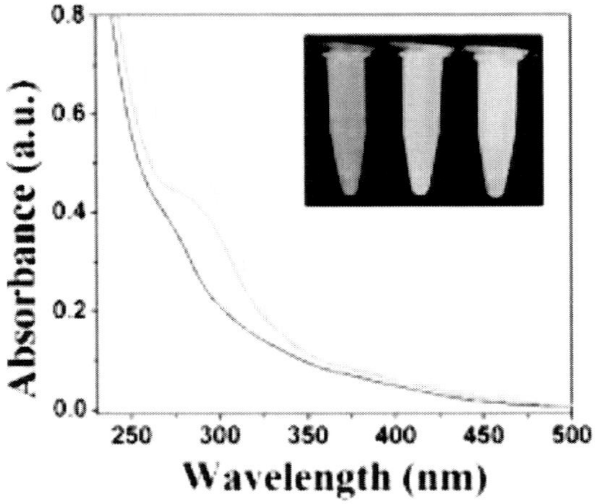

Figure 6. Ultraviolet-visible (UV-Vis) absorption spectra of Blue CDs, Green CDs and Yellow CDs in aqueous solutions. (Reproduced with permission from reference [13]. Copyright 2017 American Chemical Society).

Photo Luminescence

It is clear that using appropriate excitation wavelengths or band pass filters, different emission colors can be obtained from carbon dots dissolved in a single solvent. While some carbon dots exhibit narrow emission bands in the red near infrared (NIR) range, others emit mostly blue light or green light and have a full width at half maximum intensity in excess of 100 nm (Figure 7). The life time of carbon dots is typically in the picoseconds-nanosecond range but their quantum yield vary widely. However, there are some optical properties that are common to several structurally different members of this carbon dots family. To make the compound photoluminescent, the materials can be induced by introducing structural disorder, such as functionalization in the carbon centers surrounded by aromatic domains. Such a process can lead to a sp^3-hybridized region, thereby creating band gaps to provide photo-luminescent-active compounds [35-37].

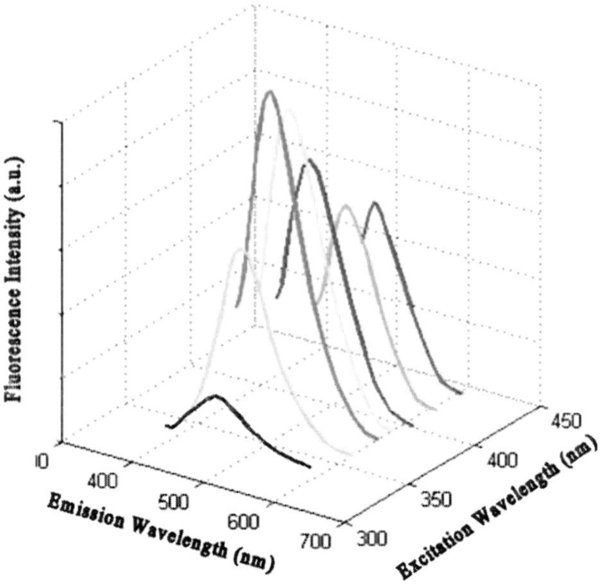

Figure 7. Photoluminescence spectra of the CDs at different excitation wavelengths. (Reproduced with permission from reference [34]. Copyright 2017 Springer Nature).

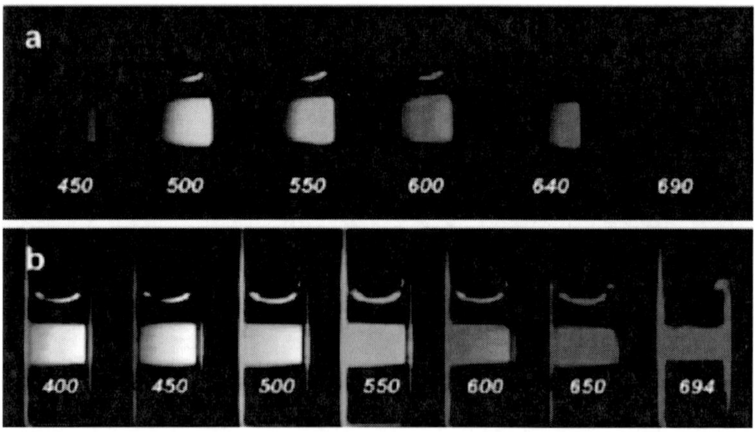

Figure 8. Aqueous solution of the PEG^{1500N}-attached CDs, (a) excited at 400 nm and photographed through band pass filters of different wavelengths as indicated; (b) Excited at the indicated wavelengths and photographed directly. Reproduced with permission from reference [7]. Copyright 2006 American Chemical Society).

The fluorescence spectra of the carbon dots' excitation wavelength vary at different wavelengths in 10-nm increments. As the excitation wavelength was set to 280 nm, the carbon dots emitted the strongest fluorescence peak at 314 nm. The maximum fluorescence wavelength of the carbon dots was gradually red-shifted as the excitation wavelength became longer (Figure 8) [2-8].

MODIFICATION AND CYTOTOXICITY ANALYSIS OF CARBON DOTS FOR SENSING AND IMAGING

Carbon dots provide a solid foundation for the development of sensors. However, carbon dots alone are not usually fluorescent and do not show biological functionality due to poor interaction with biological systems. Usually by changing the carbon dots' core size, chemical composition and structure, the photo physical properties of the compound can be controlled. For instance, water-soluble carbon dots surrounded with -COOH groups can readily combine with N-(2-aminoethyl)-N,N,N-tris(pyridin-2ylmethyl) ethane-1,2-diamine-AE-TPEA) and 4-(aminomethylphenyl)-2,2,6,2-terpy-ridine (AE-TPY), for the detection of Cu^{2+} and pH in biological systems, respectively. The π-π stacking interactions between carbon dots and DNA or aptamer can also be used as binding strategy for detection of nucleic acid. Specific receptors are needed to conjugate with carbon dots for constructing fluorescent nanosensors with high specificity. One of the simplest and most popular bio-conjugation methods is covalent bond formation through reactive functional groups such as primary amines, carboxylic acids and hydroxyls. Some carbon dots treated by acid reflux can be used directly as pH sensors due to the hydroxyl and carboxyl groups on the surface. Modification with specific coating materials may produce strong carbon dot system with controllable biocompatibility and stability [12, 23, 38-40].

Moreover, cytotoxicity studies are extremely important for the use of any material in biomedical applications. Carbon dots have been studied for their cytotoxicity effects *In vitro* and *In vivo* environments. Liu et al. reported that carbon dots prepared from candle soot by hydrothermal synthesis showed low cytotoxicity [24]. This assumption was made from the results of toxicity studies of the CDs on Hela cells performed by CKK-8 method. As shown in Figure 9, Yang et al. reported the toxicity evolutions based on the effects of carbon dots and PEG^{1500N} on proliferation, mortality and viability of human breast cancer cells (MCF-7) and human colorectal adenocarcinoma cells (HT-29) [25]. The results showed that carbon dots affected these parameters on these cells, which are comparable to those obtained from the use of $PEG^{1500N.}$ These studies also suggested that they were not imposed any significant toxic effects on mice at dosage levels even beyond that is used for *In vivo* optical imaging applications. The cell concentration and exposure time were taken more than that used for most practical biomedical applications (Figure 9). Thus, it was concluded that carbon dots should be considered as nontoxic substance [33, 40-42].

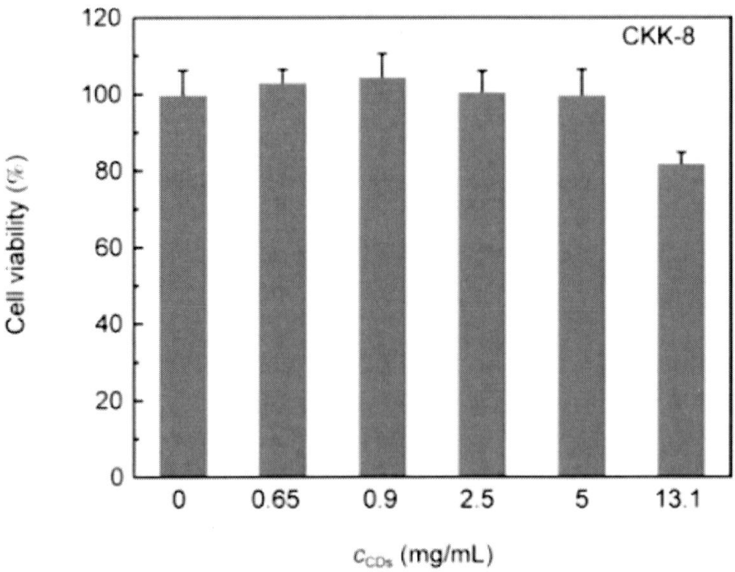

Figure 9. Cytotoxicity analysis of CDs for Hela cell performed using CKK-8 method. (Reproduced with permission from reference [25].
Copyright 2011 American Chemical Society).

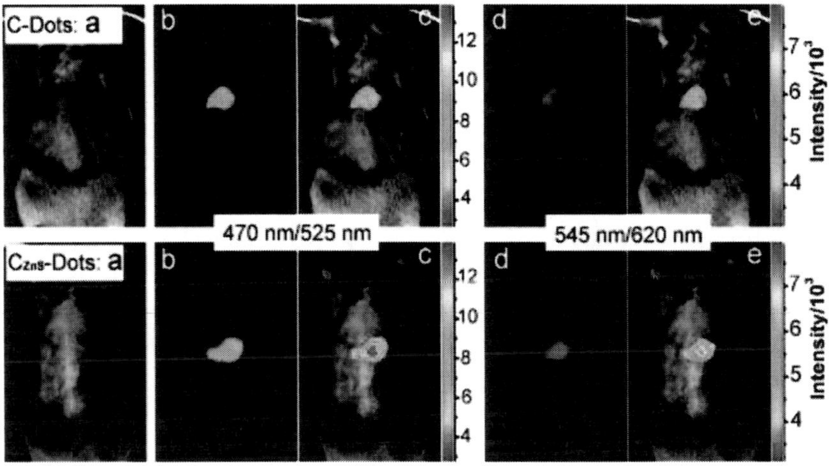

Figure 10. Subcutaneous injection of CDs (top) and ZnS doped CDs in mice (a) photograph under bright field (b,d) as detected fluorescence (excitation/emission included), (c,e) color coded images. (Reproduced with permission from reference [3]. Copyright 2009 American Chemical Society).

Using the microwave synthetic method, Wang et al. prepared carbon dots that were shown to readily enter E coli and 293 cells *In vitro,* showing multicolor luminescence that have potential use for biolabelling and bioimaging. The first reported *In vivo* study of carbon dots for optical imaging in live mice was done by Yang et al., where the carbon dots acted as bright fluorescent and contrast agent [3]. Carbon dots and ZnS doped carbon dots with PEG^{1500N} are used as passivating agents (Figure 10). As revealed from the experimental results on mice, no toxicological effect was observed, implying an excellent biocompatibility and non toxic nature of carbon dots for imaging and other biological applications [3-5, 41-43].

Flourescent images of the carbon dots synthesized in small size through soot based approach in live cells can be obtained by incubating with EAC cells only for 30 minutes. The imaging results show that cells turn bright bluish green under UV excitation and yellow under blue excitation, but colorless without uptake of carbon dots (Figure 11). This suggests that carbon dots enter into the live cells, and are available for fluorescent bioimaging of live cells [8, 38-44].

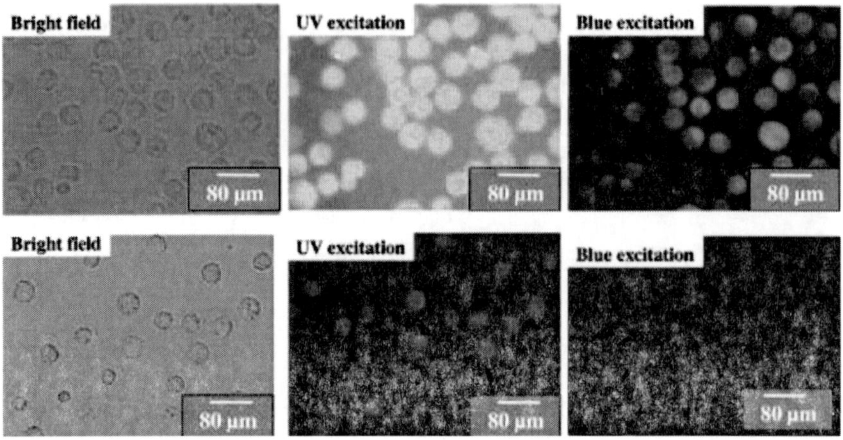

Figure 11. Top: Labelling of EAC cells using carbon dots under bright field, UV excitation and blue excitation. Bottom: Corresponding images without carbon dots. (Reproduced with permission from reference [8].
Copyright 2006 American Chemical Society).

CANCER IMAGING BY USING MULTICOLOR CARBON DOTS

Fluorescent carbon dots with the high biocompatibility and exceptionally low *in vivo* toxicity have shown great potential uses in biological systems. Carbon dots have large surface area, porous structure, non specific binding site and low toxicity. Therefore, carbon based nanomaterials have been explored for highly effective and economical transmembrane drug delivery carriers that can release drug molecules inside the tumor cells for effective cancer treatment. The exogenous site targeting and responsive drug release strategy using an external stimulus such as magnetic field or near infrared (NIR) light is a great method for *in vivo* applications because they do not need to change the specific physical or chemical properties (pH, temperature) of the environmental medium, which may cause severe side-effects to normal cells and tissues. Moreover, magnetic-functionalized carbon-based drug carrier can trigger the release of loaded drug by a simple exposure to an appropriate alternating magnetic field benefited from the hyperthermia of magnetic NPs in alternating magnetic field [45-47].

On the other hand, a NIR-responsive drug carrier brings a new opportunity to release the loaded drug at a desired area and time by applying NIR light on the tumor site, which could be improved the therapeutic efficacy in cancer treatment while minimizing the side effects. NIR light has been widely used as an external stimulus to control the drug release because of their low energy absorption, deep penetration and minimum side effects for cells and tissues. Carbon based drug carriers functionalized with inorganic nano-materials are efficiently used to convert NIR photons to heat, that is the primary technique for NIR responsive drug release. In the meantime, the elevated temperature of the NIIR responsive drug carrier under the NIR beam could synergistically kill the tumor cells, thereby improving the therapeutic efficiency. Most of the biomedical techniques used to detect cancer cells are based on the use of specific ligands for protein recognition. For example, aptamers and proteins including antibodies and enzymes, have been used for the detection of cancer cells, because of their specificity and binding affinity [48-50].

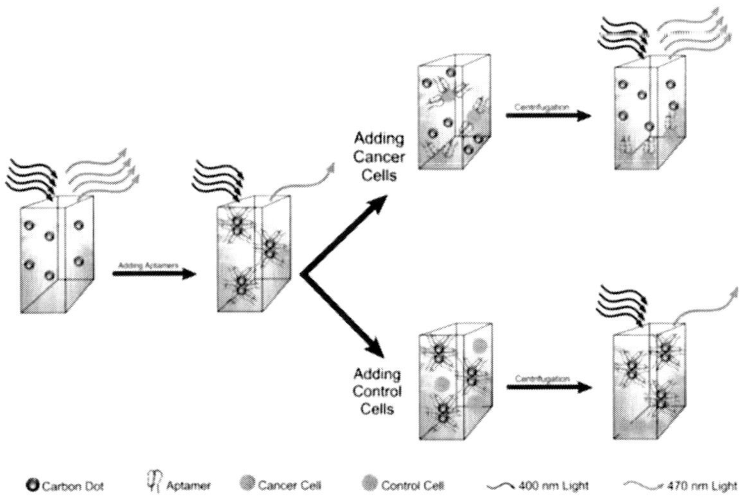

Figure 12. Schematic representation of the principle of cancer cell detection using the carbon dots-AS1411 aptamer nanoconjugates. (Reproduced with permission from reference [34]. Copyright 2017 Springer Nature).

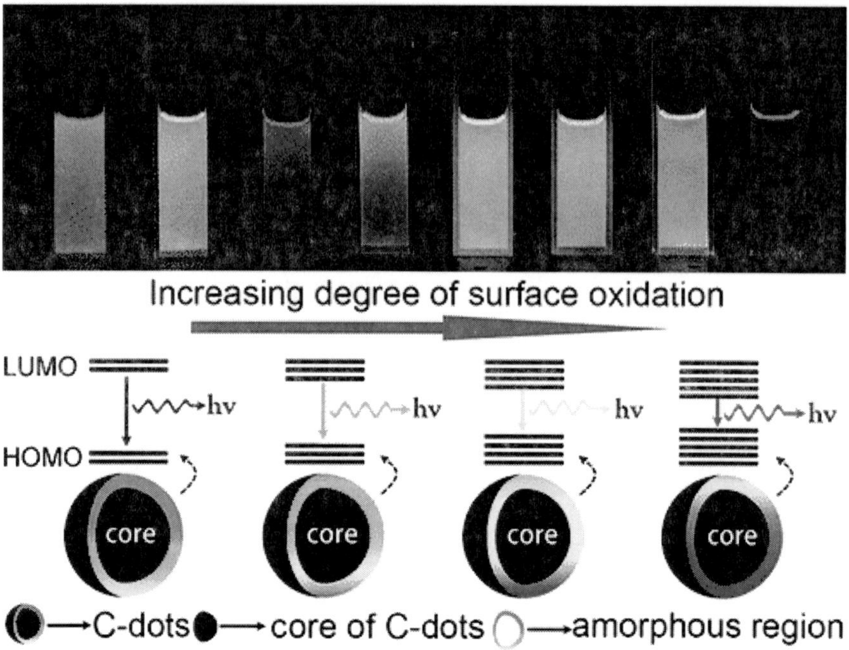

Figure 13. Schematic representation of CDs with selected precursors. (Reproduced with permission from reference [36]. Copyright 2016 American Chemical Society).

Aptamers are known for alternatives to antibodies in protein recognition and sensing, for their straight-forward synthesis, excellent controllability and broad applications. They can bind various targets, such as inorganic ions, small molecules, proteins, and even whole cells. Motaghi et al. showed a sensitive and selective signal-on method for aptamer based spectrofluorometric detection of cancer cells [34]. A nucleolin aptamer or nucleolin known as AS1411, is a multifunctional protein located primarily in the nucleolus, but is also found in the cytoplasm and on the membrane of cells. An AS1411 aptamer binds to nucleolin with high affinity. An AS1411 is wrapped around water-soluble carbon dots and used as a probe for the detection of several types of cancer cells. This protein is over-expressed in many types of tumor cells compared to normal cells, and cancer cells consequently display a higher amount of nucleolin on their surface. The Motaghi group selected mouse breast 4T1, human breast MCF7, and human cervical HeLa cancer cells as target cells, while human foreskin fibroblast cells HFFF-PI6 served as control cells (Figure 12). For the sensitive and selective spectrofluorimetric detection of target cancer cells in the presence of control cells, the cells were incubated in carbon dots-aptamer solutions. The cell suspensions were subsequently centrifuged and the fluorescence intensities were measured as an analytical signal. The specific targeting of cancer cells by AS1411 aptamers causes the release of carbon dots and enhances the fluorescence intensity. In addition, no significant change in the signal was detected by modifying the amount of human foreskin fibroblast control cells [34-39].

Ding et al. performed an experiment where carbon nanoparticles become photoluminescent upon surface passivation with oligomeric polymer chains [36]. In this work, the dependence of the carbon dots photoluminescent properties on the passivation polymer selection has been demonstrated by conjugating polyethylene glycol (PEG) chains, polyethylenimide-co-polyethylene glycol-co-polyethylenimide copolymer, and armed PEG molecules, respectively. The cytotoxicity and cellular internalization of the resulting three types of photoluminescent nanoformulations of carbon dots, named CD2, CD3, and CD4, were evaluated (Figure 13). These nano-formulations exhibited no apparent cytotoxicity on their own and were shown to successfully target cancer cells by conjugation with transferring [47-51].

A number of researchers synthesized biocompatible fluorescent carbon dots by a one-step hydrothermal method using potato as a raw material [51]. They found that the carbonization of potato at ~170°C for 12 hours produced highly fluorescent carbon dots of 0.2–2.2 nm size. The synthesized carbon dots are well dispersed in water and exhibited strong blue and bright blue emissions under UV illumination (λ_{ex} = 302 and 365 nm). The carbon dots showed a strong emission peak at 455 nm at an excitation wavelength of 374 nm. The carbon dots acted as fluorescent probes for multicolor (blue, green and red) imaging of HeLa cells and the carbon dots did not induce cell death, which indicates that the carbon dots are biocompatible and nontoxic to HeLa cells (Figure 14). Therefore, the carbon dots can be used as probes for cell-imaging applications *In vitro* and *in vivo* [45-51].

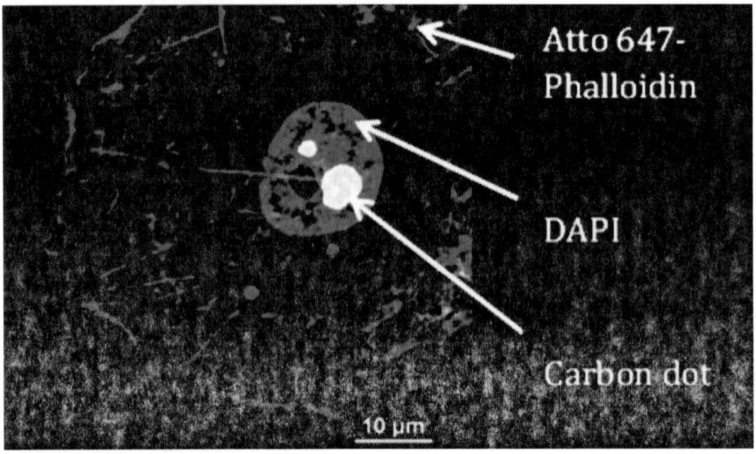

Figure 14. Cell imaging with multicolor CDs. (Reproduced with permission from reference [51]. Copyright 2018 American Chemical Society).

Bacteria and Virus Imaging Using Multicolor Carbon Dots

Currently, nano materials have sprung up as fresh antiviral agents because of their unique chemical and physical properties. Carbon dots with various surface passivations have been studied extensively over the past few years in fluorescence based live cell assays. Disease-related biomarkers can be rapidly and directly detected by carbon dots through an accurate process and considerably higher sensitivity. Not only does the nanotechnology allow the early identification of viruses, bacteria, and circulating tumor cells, it also has proven potential for single-cell analysis. The size and properties of nanoparticles make them excellent platforms for the detection and identification of pathogens in native biological samples [52-54].

One research group developed green fluorescent carbon dots that have been synthesized from a green source, gram, without any sort of covalent or ionic modifications. These gram-derived carbon dots are unique with respect to synthetic commercial a cell-tracking dye as they are non-toxic, cell internalization occurs quickly, and they have excellent bioconjugation with bacterial cells. Another potential application has been proposed, by the Sun group where carbon dots were synthesized via Laser abalation of carbon targets followed by HNO_3 reflux and passivation with PEG^{1500N} [7]. No detectable fluorescence was observed for raw carbon dots or after refluxing with acid. However, a bright luminescent was clearly visible after surface passivation with PEG^{1500N} (Figure 15). Confocal microscopy images of E. coli ATCC 25922 labeled with the functionalized carbon dots were obtained at different excitation wavelengths [7, 52-55].

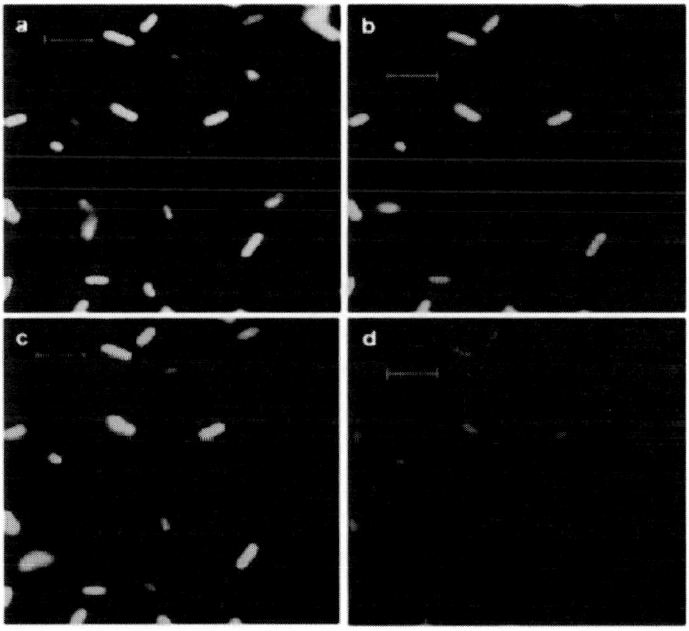

Figure 15. Confocal microscopy images of E. coli ATCC 25922 cells labeled with the CDs (a) λ_{ex} = 458 nm, detected with 475 nm long pass filter (b) λex = 477 nm, detected with 505 nm long pass filter; (c) λ_{ex}= 488 nm, detected with 530 nm long pass filter; and (d) λ_{ex} = 514 nm, detected with 560 nm long pass filter. (Reproduced with permission from reference [7]. Copyright 2006 American Chemical Society).

Essential oil from peppermint plants was used to prepare luminescent carbon nanoparticles via a simple, one-step, thermal synthesis process [53]. The peppermint oil nanoparticles had a narrow particle size distribution (1.5 ± 0.5 nm) with prominent blue emission under UV irradiation.

Photoluminescence (PL) spectra of the peppermint oil NPs dispersion exhibited characteristic emission peaks when excited from 350 to 540 nm (Figure 16). Transmission Fourier transform infrared (FTIR) spectra and X-ray photoelectron spectroscopy (XPS) were used to characterize the NP chemical composition, and revealed that functional groups, such as C═O, C–O, and −CH, were present on the nano particle surfaces, which could act as fluorescent emissive traps. Additionally, the nano particles exhibited strong antimicrobial efficiency and demonstrated good fluorescent emission during bacteria imaging, making them good candidates as antibacterial agents and multifluorescence tracers for bacterial disease treatment [52-53].

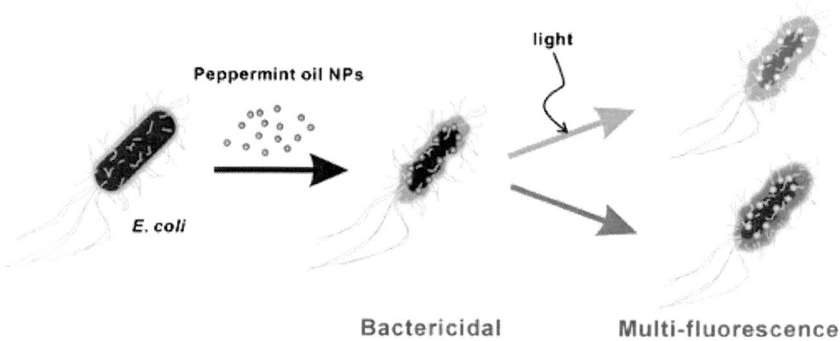

Figure 16. Bacteria imaging with Multi color NPs. (Reproduced with permission from reference [53]. Copyright 2014 American Chemical Society).

Brain Imaging Using Multicolor Carbon Dots

Fluorescent carbon dots were evaluated for their cytotoxicity and *In vivo* toxicity and also for their optical imaging performance, the results suggested that the carbon dots were biocompatible, and their performance as fluorescence imaging agents was competitive. The implication to the use of carbon dots for *In vitro* and *In vivo* application is still an ongoing research project for scientists [45-48].

A new type of carbon dots (CD-Asp) with targeting function toward brain cancer glioma was synthesized via a straightforward pyrolysis route by using D-glucose and L-aspartic acid as starting materials [57]. The synthesized CD-Asp exhibits not only excellent biocompatibility and tunable full-color emission, but also significant capability of targeting C6 glioma cells without the aid of any extra targeting molecules (Figure 17). *In vivo* fluorescence images showed high-contrast bio-distribution of CD-Asp 15 min after tail vein injection. A much stronger fluorescent signal was detected in the glioma site than that in normal brain, indicating their ability to freely penetrate the blood–brain barrier and precisely targeting glioma tissue. However, its counterparts, the carbon dots synthesized from D-glucose (CD-G), L-asparic acid (CD-A), or D-glucose and L-glutamic acid (CD-Glu) have no or low selectivity for glioma. Therefore, CD-Asp could act as a fluorescence imaging and targeting agent for noninvasive glioma diagnosis. This work highlights the potential application of CDs for constructing an intelligent nanomedicine with integration of diagnostic, targeting, and therapeutic functions [50, 56-57].

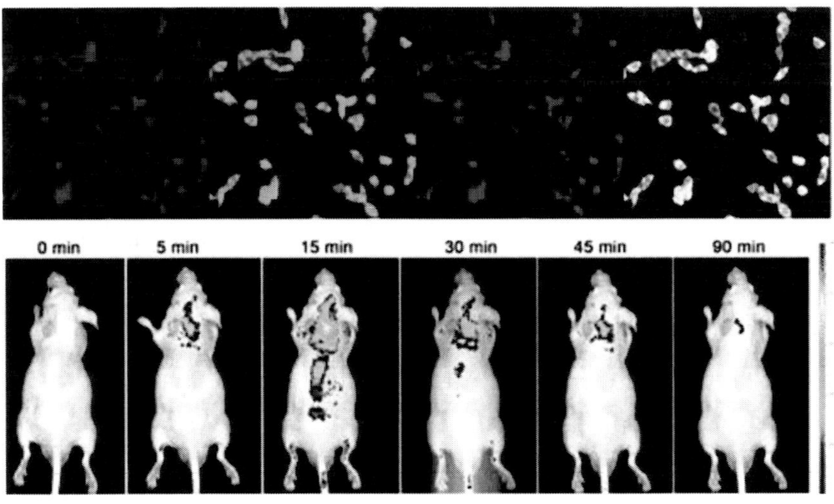

Figure 17. Self targeting fluorescent CDs for diagnosis of Brain cancer cells. (Reproduced with permission from reference [57]. Copyright 2015 American Chemical Society).

Conclusion and Future Outlook

In this chapter, we focused on multi-color carbon dots, their characterization and properties, modification for cell line and bioimaging applications in cancer cells, virus, bacteria and brain imaging. Carbon dots are newly developing fluorescent materials, which possess the potential for replacing the traditional fluorescent materials. They are expected to be great materials for *In vivo* and *In vitro* applications, because of their good aqueous solubility, bright photoluminescent characteristics, photochemical stability, low toxicity as well as biocompatibility. The carbon dots with long wavelength excitation and two photon absorption exhibit good performance in the bioimaging field. Especially the photoluminescent feature is extremely important for bioimaging and sensing. Furthermore, carbon dots have excellent chemical compositions forming active surface that can be passivated with other compounds. As a result the carbon dots based composites could have the integrated advantage of targeting in cellular imaging and related biomedical applications. The abundant species of carbon dots and their simple and easy synthesis procedures show potential for diagnostic and therapeutic applications *In vivo* imaging. However, there are some drawbacks such as most reported carbon dots can only be excited by UV light, limiting their use in *In vivo* imaging. For more effective results in imaging carbon dots with high quantum yield and NIR emission are highly desired. Since the discovery of carbon dots and analyzing their wide range of applications and properties that have been reported during the last decade, it can be stated that these luminescent and biocompatible carbon dots can come forth as novel probe for many applications in future.

References

[1] Wang, J., Qiu, J. (2016). A Review of Carbon Dots in Biological Applications, *J. Mater. Sci.* 4, 1–11.

[2] Koza, K. Ondrei., Sudolska, M., Pramanik, G., Cigler, P., Oteypka, M., Zboril, R. (2016). Photoluminescent Carbon Nanostructures. *Chem. Mater*. 28, 4085−4128.

[3] Yang, S. T., Wang, X., Wang, H., Lu, Fushen., Luo, P. G., Cao, L., Meziani, M. J., Liu, J. H., Liu, Y., Chen, M., Huang, Y., Sun, Y. P. (2009). Carbon Dots as Nontoxic and High Performance Fluorescence Imaging Agents. *J. Phys. Chem. C*. 113, 18110–18114.

[4] Sun, Y. P., Wang, X., Lu, F. S., Cao, L., Meziani, M. J., Luo, P. J. G., Gu, L. R., Veca, L. M. (2008). Doped Carbon Nanopaticles as a new Platform for Highly Photoluminescent Dots. *J. Phys. Chem*. C. 112, 18295–18298.

[5] Chandra, S., Das, P., Bag, S., Laha, D.; Pramanik, P. (2011). Synthesis, Functionalization and Bioimaging Applications of Highly Fluorescent Carbon Nanoparticles. *Nanoscale*. 3, 1533–1540.

[6] Xu, X., Ray, R., Gu, Y., Ploehn, H. J., Gearheart, L., Raker, K., Scrivens, W. A. (2004). Electrophoretic Analysis and Purification of Fluorescent Single walled Carbon Nanotube Fragments. *J. Am. Chem. Soc*. 126, 12736–12737.

[7] Sun, Y. P., Zhou, B., Lin, Y., Wang, W., Fernando, K. A. S., Pathak, P., Meziani, M. J., Harruff, B. A., Wang, X., Wang, H. (2006). Quantum-Sized Carbon Dots for Bright and Colorful Photoluminescence. *J. Am. Chem. Soc*. 128, 24, 7756−7757.

[8] Ray, S. C., Saha, A., Jana, N. R., Sarkar, R. (2009). Fluorescent Carbon Nanoparticles: Synthesis, Characterization, and Bioimaging Application. *J. Phy. Chem. C*. 113, 18546–1855.

[9] Lim, S. Y., Shen, W., Gao, Z. (2015). Carbon Quantum Dots and their Applications. *Chem. Soc. Rev*. 44, 362−381.

[10] Wang, R., Lu, K. Q., Tang, Z. R., Xua, Y. J. (2017). Recent Progress in Carbon Quantum Dots: Synthesis, Properties and Applications in Photocatalysis. *J. Mater. Chem. A*. 5, 3717–3734.

[11] Sun, X., Lei, Y. (2017). Fluorescent Carbon Dots and Their Sensing Applications. *Trends Anal. Chem.* 89, 163–180.

[12] Song, Y., Zhu, S., Yang, B. (2014). Bioimaging Based on Fluorescent Carbon Dots, *RSC Advances*. 4, 27184–27188.

[13] Wang, T. Y.; Chen, C. Y.; Liao, W. S.; Multicolor Functional Carbon Dots via One-Step Refluxing Synthesis, *ACS Sens.*, 2017, *2*, 354–363.

[14] Bourlinos, A. B., Stassinopoulos, A., Anglos, D., Zboril, R., Georgakilas, V., Giannelis, E. P. (2008). Photoluminescent Carbogenic Dots. *Chem. Mater.* 20, 4539–4541.

[15] Sahu, P. G., Yang, S. T., Sonkar, S. K., Wang, J., Wang, H., LeCroy, G. E., Cao, L., Sun,Y. P. (2013). Carbon Quantum Dots for Optical Bioimaging. *J. Mater. Chem. B*. 1, 2116–2127.

[16] Hu, S. L., Niu, K. Y., Sun, J., Yang, J., Zhao, N. Q., Du, X. W. (2009). One-Step Synthesis of Fluorescent Carbon Nanoparticles by Laser Irradiation. *J. Mater. Chem.* 19, 484–488.

[17] Zhu, H., Wang, X. L., Li, Y. L., Wang, Z. J., Yang, F., Yang, X.(2009). Microwave Synthesis of Fluorescent Carbon Nanoparticles with Electro-Chemiluminescence Properties. *Chem. Commun.* 14, 5118–5120.

[18] Li, W., Zhang, Z., Kong, B., Feng, S., Wang, J., Wang, L., Yang, J., Zhang, F., Wu, P., Zhao, D. (2013). Simple and Green Synthesis of Nitrogen-Doped Photoluminescent Carbonaceous Nanospheres for Bioimaging. *Angew. Chem. Int. Ed.* 52, 8151–8155.

[19] Zhao, S., Lan, M., Zhu, X., Xue, H., Ng, T. W., Meng, X., Lee, C-S., Wang, P., Zhang, W. (2015). Green Synthesis of Bifunctional Fluorescent Carbon Dots from Garlic for Cellular Imaging and Free Radical Scavenging. *ACS Appl. Mater. Interfaces*. 7, 17054–17060.

[20] Wang, T. Y., Chen, C. Y., Wang, C. M., Tan, Y. Z., Liao, W. (2016). Multicolor Functional Carbon Dots via One Step Refluxing Synthesis. *ACS Sens*. 10, 1021–1025.

[21] Xu, Y., Wu, M., Liu, Y., Feng, X. Z., Yin, X. B., He, X. W., Zhang, Y. K. (2013). Nitrogen-Doped Carbon Dots: A Facile and General Preparation Method, Photoluminescence Investigation, and Imaging Applications. *Chem. Eur. J.* 19, 2276–2283.

[22] Li, C. X., Yu, C., Wang, C. F., Chen, S. (2013). Facile plasma-induced fabrication of fluorescent carbon dots toward high-performance white LEDs. *J. Mater. Sci.* 48, 6307–6311.

[23] Zhu, S., Meng, Q., Wang, L., Zhang, J., Song, Y., Jin, H., Zhang, K., Sun, H., Wang, H., Yang, B. (2013). Highly Photoluminescent Carbon Dots for Multicolor Patterning, Sensors, and Bioimaging. *Angew. Chem. Int. Ed.* 52, 3953–3957.

[24] Liu, H., Ye, T., Mao, C. (2007). Fluorescent Carbon Nanoparticles Derived from Candle Soot. *Angew. Chem. Int. Ed.* 46, 6473–6475.

[25] Yang, S., Cao, L., Luo, P., Lu, F., Wang, X., Wang, H., Meziani, M., Liu, Y., Qi, G., Sun, Y. (2009). Carbon Dots for Optical Imaging *In vivo*. *J. Am. Chem. Soc.* 131, 11308–11309.

[26] Peng, H., Travas-Sejdic, J. (2009) Simple Aqueous Solution Route to Luminescent Carbogenic Dots from Carbohydrates, *Chem. Mater.* 21, 5563–5565.

[27] Liu, L., Li, Y., Zhan, L., Liu, Y., Huang, C. (2011). One Step Synthesis of Fluorescent Hydroxyls Coated Carbon Dots with Hydrothermal Reaction and its Application to Optical Sensing of Metal Ions. *Science China Chemistry*. 54, 1342–1347.

[28] Jiang, K., Sun, S., Zhang, L., Lu, Y., Wu, A., Cai, C. (2015). Red, Green, and Blue Luminescence by Carbon Dots: Full-Color Emission Tuning and Multicolor Cellular Imaging. *Angew. Chem. Int.* 54, 5360–5363.

[29] Wang, F., Pang, S., Wang, L., Kreiter, Q. L. M., Liu, C. Y. (2010). One Step Synthesis of Highly Luminescent Carbon Dots in Non-coordinating Solvents. *Chem. Mater.* 22, 4528–4530.

[30] Sahu, S., Behera, B., Maiti, T. K., Mohapatra, S. (2012). Simple One-step Synthesis of Highly Luminescent Carbon Dots from Orange Juice: Application as Excellent Bioimaging Agents. *Chem. Commun.* 48, 8835–8837.

[31] Wang, X., Qu, K., Xu, B., Ren, J., Qu, X. (2011). Microwave Assisted One-step Green Synthesis of Cell Permeable Multicolor Photoluminescent Carbon Dots Without Surface Passivation Reagents. *J. Mater. Chem.* 21, 2445–2450.

[32] Armaković, S., Armaković, S. J., Koziel, S. (2017). Optoelectronic Properties of Curved Carbon Systems. *Carbon.* 111, 371–379.

[33] Wang, Y., Anilkumar, P., Cao, L., Liu, J. H., Luo, P. G., Tackett II, K. N., Sahu, S., Wang, P., Wang, X., Sun, Y-P. (2011). Carbon Dots of Different Composition and Surface Functionalization: Cytotoxicity Issues Relevant to Fluorescence Cell Imaging. *Exp. Biol. Med.* 236, 1231–1238.

[34] Motaghi, H., Mehrgardi, M. A., Bouvet, P. (2017). Carbon Dots-AS1411 Aptamer Nanoconjugate for Ultrasensitive Spectro-fluorometric Detection of Cancer Cells. *Scientific Reports*. 7, 10513.

[35] Chen, T. H., Tseng, W. L. (2017). Self-Assembly of Monodisperse Carbon Dots into High-Brightness Nanoaggregates for Cellular Uptake Imaging and Iron (III) Sensing. *Anal. Chem. 89,* 11348–11356.

[36] Ding, H., Yu, S. B., Wei, J. S., Xiong, H. M. (2016). Full-Color Light-Emitting Carbon Dots with a Surface-State-Controlled Luminescence Mechanism. *ACS Nano*. 10, 484–491.

[37] Qian, Z., Ma, J., Shan, X., Feng, H., Shao, L., Chen, J. (2014). Highly Luminescent N-Doped Carbon Quantum Dots as an Effective Multifunctional Fluorescence Sensing Platform. *Chem. Eur. J.* 20, 2254–2263.

[38] Igor, L., Uyeda, H. T., Goldman, E. R., Mattoussi, H. (2005). Quantum Dot Bioconjugates for Imaging, Labelling and Sensing. *Nat. Mater*. 4, 435–446.

[39] Liu, J. H., Cao, L., Le Croy, G. E., Wang, P., Meziani, M. J., Dong, Y., Liu, Y., Luo, P. G., Sun, Y. P. (2015). Carbon Quantum Dots for Fluorescence Labeling of Cells, *ACS Appl. Mater. Interfaces*. 7, 19439–19445.

[40] Ding, C., Zhu, A., Tian, Y. (2014). Functional Surface Engineering of Carbon Dots for Fluorescent Biosensing and *in vivo* Bioimaging. *Acc. Chem. Res.* 47, 20–30.

[41] Lemenager, G., De Luca, E., Sun, Y. P., Pompa, P. P. (2014). Super-resolution Fluorescence Imaging of Biocompatible Carbon Dots. *Nanoscale*. 6, 8617–8623.

[42] Cao, L., Wang, X., Meziani, M., Lu, F., Wang, H., Luo, P., Lin, Y., Harruff, B. A., Veca, L. M., Murray, D., Xie, S., Sun, Y. (2007).

Carbon Dots for Multiphoton Bioimaging. *J. Am. Chem. Soc.* 129, 11318–11319.

[43] Dui, F., Li, J., Huai, Y., Zhang, M., Zhou, Z., Yuan, J., Wang, J., Peng, W., Zhang, L., Xia, S., Wang, D., Yang, S., Xu, W., Gong, A., Shao, Q. (2015). Multicolor Nitrogen doped Carbon dots for Live Cell Imaging. *J. Biomed. Nanotechnol.* 11, 780–788.

[44] Song, Y., Zhu, C., Song, J., Li, H., Du, D., Lin, Y. (2017). Drug-Derived Bright and Color-Tunable N-Doped Carbon Dots for Cell Imaging and Sensitive Detection of Fe^{3+} in Living Cells. *ACS Appl. Mater. Interfaces*. 9, 7399–7405.

[45] Feng, T., Ai, X.; An, G., Yang, P., Zhao, Y. (2016). Charge-Convertible Carbon Dots for Imaging-Guided Drug Delivery with Enhanced *in vivo* Cancer Therapeutic Efficiency. *ACS Nano*. 10, 4410–4420.

[46] Pardo, J., Peng, Z., Leblanc, R. M. (2018). Cancer Targeting and Drug Delivery Using Carbon-Based Quantum Dots and Nanotubes, *Molecules*. 23, 378.

[47] Wang, Q., Huang, X., Long, Y., Wang, X., Zhang, H., Zhu, R., Liang, L., Teng, P., Zheng, H. (2013). Hollow Luminescent Carbon Dots for Drug Delivery, *Carbon*. 59, 192–199.

[48] Wang, H., Cao, G., Gai, Z., Hong, K., Banerjee, P., Zhou, S. (2015). Gold Capped Magnetic Fluorescent Carbon Hybrid Nanoparticles for Multicolor Cell Imaging, Magnetic/NIR-Responsive Drug Carrier and Enhanced Photo-thermo Therapy. *Nanoscale.* 7, 7885–7895.

[49] Liu, Z., Liang, X. J. (2012). Nano Carbons as Theranostics. *Theranostics*. 2, 235–237.

[50] Zhang, H., Yee, D., Wang, C., (2008). Quantum Dots for Cancer Diagnosis and Therapy: Biological and Clinical Perspectives. *Nanomedicine*. 3, 83–91.

[51] Syamantak Khan, Navneet C. Verma, Chethana, and Chayan K. Nandi, Carbon Dots for Single-Molecule Imaging of the Nucleolus, *ACS Appl. Nano Mater.*, 2018, 1, 483–487.

[52] Matea, C. T., Pop, T., Mosteanu, O., Buzoianu, A. D., Puia, C., Iancu, C., Mocan, L. (2017). Development of nanoparticle based

optical sensor for pathogenic Bacterial detection. *J. Nano-biotecnology*. 15:35.

[53] Kung, M. L., Lin, P. Y., Hsieht, C. W., Tai, M-H., Wu, D. C., Kuo, C. H., Hsieh, S. L., Chen, H. T., Hsieh, S. B. (2014). Bifunctional Peppermint oil Nanoparticles for Antibacterial Activity and Fluorescence Imaging. *ACS Sustainable Chem. Eng. 2,* 1769–1775.

[54] Baig, M. M. F., Chen, Y. C. (2017). Bright Carbon Dots as Fluorescence Sensing Agents for Bacteria and Curcumin. *J Colloid Interface Sci*. 501, 341–349.

[55] Du, T., Liang, J., Dong, N., Liu, L., Fang, L., Xiao, S., Han, H. (2016). Carbon Dots as Inhibitors of Virus by Activation of Type I Interferon Response. *Carbon*. 110, 278–285.

[56] Li, Q., Ohulchanskyy, T. Y., Liu, R., Koynov, K., Wu, D., Best, A., Kumar, R., Bonoiu, A., Prasad, P. N. (2010). Photoluminescent Carbon Dots as Biocompatible Nanoprobes for Targeting Cancer Cells i*n vitro*. *J. Phys. Chem. C. 114,* 28, 12062–12068.

[57] Zheng, M., Ruan, S., Liu, S., Sun, T., Qu, D., Zhao, H. (2015). Self-Targeting Fluorescent Carbon Dots for Diagnosis of Brain Cancer Cells, *ACS Nano*. 9, 1455–1146.

In: Nanotechnology
Editor: Paresh Chandra Ray
ISBN: 978-1-53613-889-4

Chapter 7

NANOCEUTICALS: POTENTIAL APPLICATIONS FOR MITIGATING NUTRITIONAL DEFICIENCIES

Anu Kumar[1], Nikhil Kumar[2], Tejpal Dhewa[3] and Francesco Marotta[4,*]

[1]University Institute of Biotechnology, Chandigarh University, Gharuan, Mohali, Punjab, India

[2]Faculty of Applied Sciences, Shri Venkateshwara University, JP Nagar, Uttar Pradesh, India

[3]Department of Nutrition Biology, School of Interdisciplinary and Applied Sciences, Central University of Haryana, Mahendergarh, Haryana, India

[4]ReGenera R&D International for Aging Intervention, Milano-Beijing, Italy-China & VCC Preventive Medical Promotion Foundation, Beijing, China

---

[*]Corresponding Author Email: fmarchimede@libero.it.

Abstract

It is always an issue to get reliable and sustainable food sources and to keep the food fresh longer for getting the maximum benefits from them due to the increasing human population, industries and technology. Thus, there is an urgent need for innovative solutions to supply the foods with nutritional values. In this sense, use of nanotechnology is very importantand it is considered as an enabling technology to contribute in food technology. There are various nanosystems including quantum dots, nanoparticles, nanofibres, nanocapsules or nanoemulsions which are designed to play the role for developing better methods to enhance the food quality, quantity and nutritious value. However, it is always an issue of debate to cope with the possible disadvantages of such systems since technology brings the problems along with its solutions. The concept of "nanoceuticals" is gaining popularity and commercial dairy/food and food supplements containing nanoparticles are available in the market. This review chapter provides an insight of applications of nanotechnology in the field of nutrition and will shortly depicts its advantages, possible unfavorable outcomes and solutions.

Keywords: nanoparticles, nanotechnology, nanoceuticals

Introduction

Micronutrients are the non-energy yielding class of nutrients such as vitamins and minerals that arerequired in small amounts, milligrams or micrograms daily, for ensuring optimal health and for standard physiological functions. These must be supplied through diet properly in adequate amounts as these are not synthesized by our body [1]. These have enormous health significance in some biological functions such as enzyme regulation and activity of hormones, expression of genes, proliferation of cells and cell differentiation, and in immune modulation. These are necessary to metabolize and utilize macronutrients [2].

Micronutrient deficiency often remains unnoticed withinacommunity in spite of their gradual but harmful consequences on immune system functioning, growth, and mental development. Because of these reasons micronutrient deficiencies have been termed as "hidden hunger" by the scientific society. In individuals following popular diet plans, as suggested with food alone, there is high tendency of becoming micronutrient deficient; a state shown to be scientifically linked to an increased risk for many dangerous and debilitating health conditions and diseases [3].

Micronutrient Deficiency

Micronutrient deficiency is becoming aseriousconcern regarding the health of public all around the world [4, 5].It is silent outbreak of vitamin andmineral deficiencies and many people of all genders and ages are getting affected by the nutrient deficiency. Literature suggest that over 2 billion people are at risk for vitamin A, iodine, and iron deficiency globally [6]. In the previous literature survey of developing world, 20% of the population found suffering from iodinedeficiency, 25% of children have deficiency due to vitamin A, and more than 40% of women areanemic [7],increasingtheriskofdeath, morbidity and susceptibility to infection, blindness, adverse birth outcomes, stunting, decreasedcognitive capacity and mental retardation [8].

There are many specific reasons behind micronutrient deficiency such as lack of inadequate food intake. Quality of nutrients must be available properly, attributed to the presence of inhibitors, mode of preparation, and interactions, and the presence of infections [9].

Nanotechnology

In nanotechnology structures, devices or materials that have at least one dimension and is approximately 1-100 nm in length have been characterized, fabricated or manipulated [10]. Cell membranes, hormones, DNA, etc. are examples of naturally occurring Nano structures, the same as food molecules, proteins, fats, carbohydrates, etc. [10].Nanotechnology applications include a number of nano scalematerials added to foods and encapsulation techniques based on nanotechnology as delivery systems for other food components [11].Nanotechnology offerssome tools capable of manipulating the material at the nanoscale that is having the advantage of nanosize. Nano based nutrients may enhance the functionality or bioavailability of ingredients and nutrients, and thereby reduce the concentrations needed in the food product. Food companies are currently producing nanoparticles in emulsions in an attempt to control the material properties of foodstuffs, such as in the manufacture of ice cream to increase texture uniformity [12]. There are many applications of nanotechnology in every segment of the food industry (Figure 1), from agriculture to food processing to nutritional supplements. The applications of nanotechnology in the food sector are emerging day by day and are predicted to grow rapidly in the coming time. Many largest food companies in the world are reported to have been actively exploring the potential of nanotechnology for the welfare of mankind. This chapter is aimed at providing an insight into the different applications of nanotechnology in the food sector, neutraceuticals to nanoceuticals, mechanism of action of nanoceuticals and potential applications of nanoceuticals in overcoming nutritional deficiencies. No doubt neutraceuticals are available to tackle the problem of nutrition deficiency among humans, but at present,in order to address this problem development of novel, more efficient nano-based nutritional supplements, i.e.,nanoceuticals is becoming the prime research area in the food sector.

APPROACHES TO MITIGATE MICRONUTRIENT DEFICIENCY

There are some conventional interventions or strategies available to tackle the problem of micronutrient deficiency. The widespread understanding of micronutrient deficienciesaround the globe is a serious concern for public health, and the potential to address such deficiencies relatively cheaply through fortification or dietary supplements has led to a number of multidirectional efforts in favor of traditional interventions.

Supplementation and fortification are intervention strategies that often are aimed at the immediate or short-term improvement of the situation and often deal withs the symptoms of micronutrient deficiencies [13].

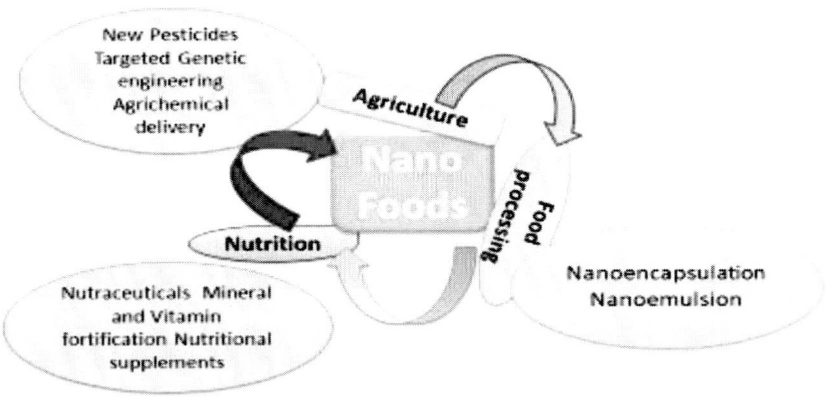

Figure 1. Applications of Nanotechnology in areas of Food science, from agriculture to nutrition and neutraceuticals.

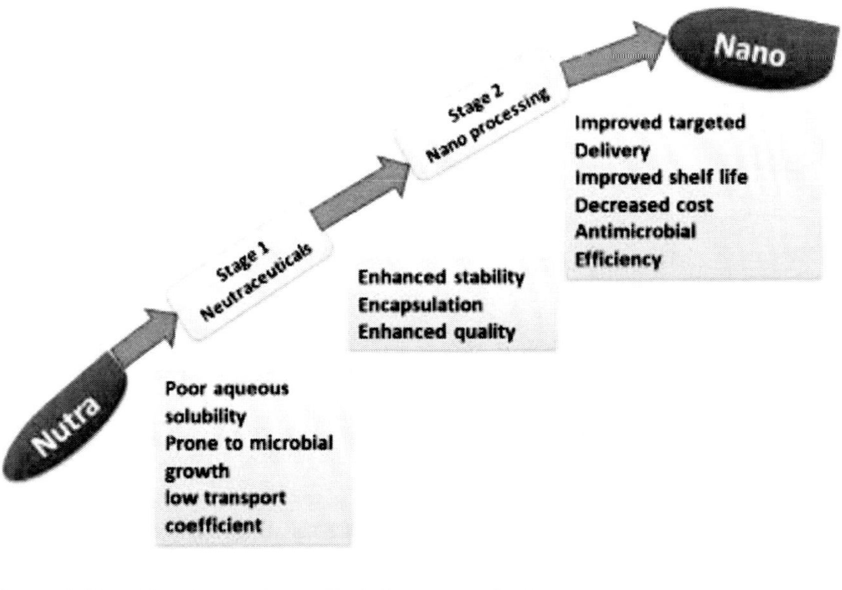

Figure 2. Transforming neutraceuticals to nanoceuticals.

In the present scenario, extensive research is being conducted to devise nano-based encapsulation techniques and specialized polymer-based or liposome-mediated has been carried out to improve the health promotion properties of different phytochemical based nutraceuticals [14]. Later improves their systemic availability, delivery properties, solubility and thus their ability to be transferred across intestinal membranes into the blood [15]. Different types of nanoceuticals including the polymeric nanoparticles, solid lipid nanoparticles, magnetic nanoparticles, metal and inorganic nanoparticles, quantum dots, polymericmicelles, phospholipid micelles, colloidal nano-liposomes, nano-emulsions dendrimers polymeric, inorganic, and hybrid nanocarriers have been prepared using techniques like high-pressure homogenization, complex coacervation, coprecipitation, salting out technique, solvent displacement method, solvent emulsification-diffusion, supercritical fluid method and self-assembly method[16].

Also, the stabilityof micronutrients was improved by using nanotechnology during processing, storage, and distribution. These modified nutraceuticals have nevertheless, reached the market and are prepared to their fullest potential in the hands of the patients/consumers.

NUTRA TO NANO

Emergence of term "Nutra to nano" is said to be a advancement and logical progression of the technology which is likely to upgradethe target specific delivery of dietary supplements,enhance bioavailability of phytochemicals, boost flavors, introducing antibacterial nanofood, enhance the shelf life of the product, and also improve the storage of food [17]. Nutritional supplements with the fusion of nanotechnology enable the drugs to be delivered more efficiently. These supplements interact most powerfully with cells and are easily acceptable. Even though nano-technology came into effect in foods by producers' it is yet to be recognized by the consumers, due to the ethical issues and unawareness [17].The genesis for this transformationis clearly depicted in Figure 2.

NANOCEUTICALS

Nanoceuticals are defined as pharmaceuticals or natural supplements also known as nutraceuticals, which are designed using "nanotechnology". "Nutraceutical", a portmanteau of the words "nutrition" and "pharmaceutical", was coined in 1989 by Stephen L. DeFelice, founder, and chairman of the Foundation of Innovation Medicine. These are the nutrients that have been manufactured in nanoparticles using nano-technology. These have a remarkable potential to solve several problems associated with dietary supplements absorption. The term is applied to products that range from isolated nutrients, dietary fiber, prebiotics, probiotics, polyunsaturated fatty acids, antioxidants and other different types of dietary supplements and herbal products, specific diets and processed foods such as cereals, soups, and beverages [18].

PROMINENT BENEFITS OF NANOCEUTICALS

Nano-processed food i.e., nanoceuticals get into food supply and overcoming nutritional deficiencies and has proved a well-defined and better and potent alternative to traditional supplements such as Dietary Supplements (Vitamins, Minerals, Herbals, Non-Herbals, & Others), and Functional Foods & Beverages [19]. Also, nanoceuticals have improved and better efficiency for overcoming nutritional deficiencies. In particular, the use of nanoceuticals is gaining much attention in this regard. Some of the advantages of nano-based food have been defined (Figure 3).

1. These ensure targeted delivery of nutrients and enhanced absorption of nutrients.
2. These are ensuring improved bioavailability and safety of nutrients in the diet.
3. Antimicrobials to combat pathogens in food and ultimately mitigating nutritional deficiency.

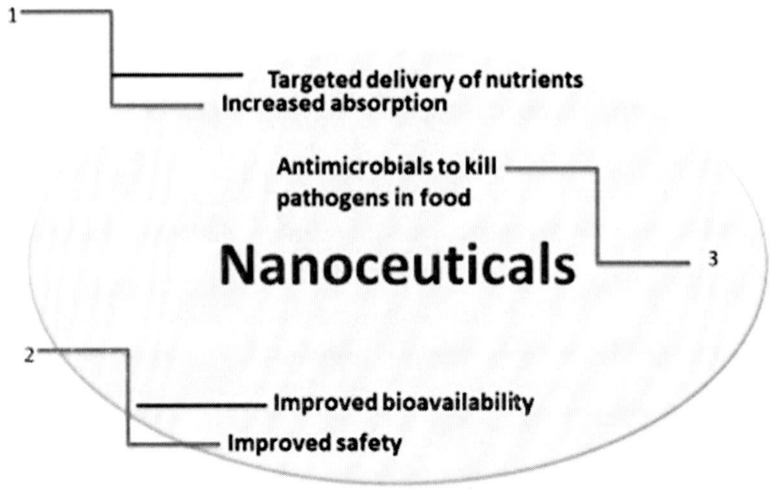

Figure 3. The most prominent benefits of nanoceuticals.

Safety Concern

Before marketing of dietary supplements it the responsibility of manufacturers of dietary supplements for ensuring the safety of their products. As these supplements consumed by humans for mitigating nutrients deficiency, the product should be safe and reliable. If the product is found to be unsafe after it hits the market, FDA is responsible for taking action against it. Experts in the industry say that the regulatory framework provided by DSHEA is adequate to ensure the safety of dietary supplements. They point out that manufacturers must notify FDA at least 75 days before marketing a dietary supplement if it contains a new dietary ingredient.

However, it is up to the manufacturer to determine whether a substance is a new dietary ingredient. "The statute really depends on the industry determining that the ingredient is new and giving notice to FDA. The company is free to determine that it is not new," and then the firm does not need to give notice to FDA, Schultz contends. The underlying issue is how

FDA defines a new dietary ingredient. According to the agency, a new substance is one that has not been previously marketed and has been chemically altered. Why nanoscale substances are not chemically altered? They are just substances reduced to the nanoscale through a physical process. Whether they should be considered new dietary ingredients it is still unclear.

However high dose consumption of nanoproducts and food supplements is associated with overconsumption and risk of toxicity. Therefore safety and efficacy of nutrients for supplementation should be addressed to avoid overconsumption of nanoproducts. However, no serious health hazard reported till now by using nutrients supplements. According to a report on Nutraceuticals Market in India, some people in the country are purchasing products including supplements, beverages, and food, designed to boost their health, prevent disease, enhance alertness and stamina [20]. Some nanoproducts are listed below in Table 1 [21].

Table 1. List of food-related nanoproducts

Name of product	Key points
Carotenoids nanoparticles	Added in fruit drinks for improved bioavailability
Canola oil based nanosized micellar system	These provide delivery of materials such as vitamins, minerals, or phytochemicals
Nanoceutical products containing nanocages or nanoclusters that act as delivery vehicles	As a sweet chocolate drink
Nanosilver or nanogold	These are available as mineral supplements

Conclusion

Altering nutraceuticals to the nano form has been noticed as a foundation stone of the forthcoming manufacturer-consumer relationship. The nanoceuticals are capable of attracting consumers because of their captivating property as a consequence of their enhanced bioavailability which has led to the oversight by the regulators which have given a green signal to abundant products which are in the market for consumer use. The regulatory agencies ought to take concrete measures to establish a standardized international board which will set guidelines, develop new policies and makes reforms to normalize the safety assessment procedure which governs the acceptance of the nanomedical products and regulate the products entering into the market. Utilization of nanoceuticals may enhance the absorption of supplements as compared to traditional dietary supplements. These have enhanced health and commercial viability and is having a very potential to mitigate nutritional deficiency. Nanoceuticals can be delivered to the desired site and also influence supplements taste, texture, improve flavor and shelf life. In the current overwhelming rush to "antiaging" or, more appropriately, "geroprotectors" compounds are aimed to design better preventive medicine policies against age-related diseases together with tentatively slow-aging interventions. In this setting nanoceuticals provide an attracting avenue and this has been extensively dealt with by Kumar and Marotta [22] in the recent much-cited book "Anti-aging Drugs".

Acknowledgments

An unbiased support for publications logistics was provided by MS Laura Bissi, PhD, CEO of San Babila Clinic, Corso Matteotti 1/A, 20121 Milano, Italy.

References

[1] Rolfes S. R., Pinna K., Whitney E. (2008). *Understanding Normaland Clinical Nutrition* (8th Ed.) Belmont: Wadsworth.

[2] Prasad K. N. (2011). *Micronutrients in Health and Disease*. Boca Raton: CRC Press.

[3] Calton J. B. (2010). Prevalence of micronutrient deficiency in popular diet plans. *J Int Soc Sports Nutr* 7: 24.

[4] Chakravarty I., Sinha R. K. (2002). Prevalence of Micronutrient Deficiency Based on Results Obtained from the National Pilot Program on Control of Micronutrient Malnutrition. *Nutr. Rev* 2: S53-S58.

[5] Uchendu F. N. (2011). Micronutrient Malnutrition, a Tragedy to Childhood Growth and Education. *Glob J Med Res* 11: 26-34.

[6] Tulchinsky T. H. (2010) Micronutrient Deficiency Conditions: Global Health Issues. *Public Health Rev* 32: 243-255.

[7] Meenakshi J. V., Johnson N. L., Manyong V. M., Degroote H., Javelosa J. et al. (2010). How Cost-Effective is Biofortification in Combating Micronutrient Malnutrition? An Exante Assessment. *World Dev* 38: 64-75.

[8] Eckhardt C. L. (2006). Micronutrient Malnutrition, Obesity, and Chronic Disease in Countries Undergoing the Nutrition Transition: Potential Links and Program/Policy Implications. *FCND Discussion Paper 213*.

[9] Ramakrishnan U. (2002). Prevalence of micronutrient malnutrition worldwide. *Nutr Rev* 60: S46-S52.

[10] Powell M., Colin M. (2008). Nanotechnology and food safety: potential benefits, possible risks. *CAB Rev Perspect Agric Vet Sci Nutr Nat Res* 3: 123-142.

[11] Nichols-Richardson S. M. (2007). Nanotechnology: Implications for Food and Nutrition professionals. *J Am Diet Assoc* 107: 1494-1497.

[12] Paul S. D., Dewangan D. (2016). Nanotechnology and Neutraceuticals. *Int J Nanomater Nanotechnol Nanomed* 2(1): 009-012.

[13] Abeshu M. A., Geleta B. (2016). The Role of fortification and supplementation in mitigating the 'hidden hunger.' *J Nutr Food Sci* 6: 459.

[14] Yoshinori, M., Eunice, L. C., Bo J. (2010). Biological active food proteins and peptides in health: an overview. In *Bioactive proteins and peptides as functional foods and nutraceuticals* 5-12.

[15] Shegokar R., Muller R. H. (2010). Nanocrystals: industrially feasible multifunctional formulation technology for poorly soluble actives. *Int J Pharm* 399(1-2):129-139.

[16] Dasgupta N., Ranjan S., Mundra S., Kumar C., Ashutosh R. (2015). Fabrication of Food grade Vitamin E nanoemulsion by low energy approach, characterization, and its application. *Int J Food Prop* 19(3):700-708.

[17] Chellaram C., Murugaboopathi G., John A. A., Sivakumar R., Ganesan S. (2014) Significance of nanotechnology in food industry *APCBEE Procedia* 8:109-113.

[18] Das L., Bhaumik E., Raychaudhuri U., Chakraborty R. (2012). Role of nutraceuticals in human health. *J Food Sci Technol* 49(2): 173-183.

[19] Smith C. (2013). *Forever Green is capitalizing on the rapidly growing nutraceutical market business wire*. OREM, Utah.

[20] Report (2011) *Exploring the role of food and drink in the future of health and wellness: an overview of our report and process*. Ireland. Retrieved from http://www.bordbia.ie/industry/manufacturers/insight/publications/bbreports/Documents/FutureofHealthand Wellness2011.pdf.

[21] Sundarraj A. A. S., Guhan Nath, Aaron S. I. and Ranganathan T. V. (2014). Recent innovations in nanotechnology in food processing and its various applications - a review. *Int J Pharm Sci Rev Res* 29(2): 116-124.

[22] Vaiserman, A. M. (Ed.). (2017). *Anti-aging drugs: from basic research to clinical practice*. Royal Society of Chemistry.

In: Nanotechnology
Editor: Paresh Chandra Ray
ISBN: 978-1-53613-889-4

Chapter 8

ETHICAL CONSIDERATIONS ON NANOTECHNOLOGY

Manuel Alberto M. Ferreira [*] ***and José António Filipe***
Instituto Universitário de Lisboa (ISCTE-IUL),
Business Research Unit (BRU-IUL) and Information Sciences,
Technologies and Architecture Research Center (ISTAR-IUL),
Lisboa, Portugal

ABSTRACT

Since a significant time ago, although time runs very fast, nanotechnology transformed from one of the most promising scientific hopes in uncountable human domains into a marvelous certainty. Innumerable scientific studies in several areas of knowledge were made since nanoscale emergence, carrying their contribution to the nanoscience development, leading to a great development of technical and scientific knowledge but also raising numerous problems in the ethical field. In this chapter, nanotechnology is discussed both in terms of ethics and in terms of borders that nanotechnology applications must satisfy and concluding notes are presented, highlighting the results of the analysis. Significant

[*] Corresponding Author Email: manuel.ferreira@iscte-iul.pt.

considerations are made on the close connection between ethics and the nanotechnology and the effects over the society and values.

Keywords: nanotechnology, society, ethics

INTRODUCTION

Bearing in mind that nanotechnology is the skill to work at the atomic, molecular and supramolecular levels (at a scale of approximately 1 – 100 nm), the enormous possibilities raised by the work of nanomaterials can be easily understood. The creation and the use of these material structures, devices and systems with essentially new properties and functions result from their very small structure, with applications in several areas, such as bioprocessing in industries; molecular medicine, considering physical, chemical, biological, and medical techniques used to describe molecular structures and mechanisms, often referred as personalized medicine; analyzing the health effect of nanostructures in the environment; improving food and agricultural systems; in summary: improving human performance at many fields, considering the different branches of science and the applications at many dimensions and scales.

Indeed nanotechnology, for instance, supplies the tools and the technology, allowing that researches lead to the transformation of biological systems, to the extent that biology provides models and bio-assembled components to nanotechnology.

Another relevant example: nanoscales, used in biosystems, contribute to enhance very innovative and promising results in medical area; improvements in telemedicine and on health may be expected with new systems operation and new nanotechniques, much less invasive and therefore much less uncomfortable for patients.

However, ethical as far as legal and social implications are postured. This discussion is as old as the issue itself but the boundaries are too much large, even large enough to find out the difficulties on finding definitive answers to this discussion. In the next section this issue is approached in general terms. Follow a section on ethical human dignity and in the next two sections are approached the themes of nanotechnology and environment, first, and of ethics and nanotechnology in the workplace. Then a framework for ethical analysis in nanotechnology is presented and this work finish with a concluding remarks section. In the end a short bibliography on these subjects is presented.

ETHICS AND NANOTECHNOLOGY

Computers have brought an increase in the ability to type and made handwriting more and more superfluous, even obsolete. Mobiles and their miniaturization brought the ability to generally communicate from anywhere with anybody, making more and more useless the traditional phones. Nanotechnology, because of the negligible dimension of devices this technology proportionate, will spread the use of devices that now are nuisance. For instance, the pacemaker that presently is used like a "last resource", certainly soon will be used in a much larger scale because it is hopeful that will be possible to design it at a nanoscale level.

However ethical questions quickly rise with the development of nanotechnology:

- What are the borders for this kind of progress?
- Until where it is admissible and wise to go?
- What are the implications in a later phase of development – obviously unknown - of these technologies?

Global discussion has begun about this theme since the beginning of the discoveries about this subject. Legal, ethical and social implications are irreversible and the discussion is usually asked to be hard and not pacific, needing urgently great clarifications.

Vanessa Nurock - see citation in Bensaude-Vincent (2010)- questions the standard view of an ethics for nanotechnology. She argues that none of the current trends in the discipline of ethics would qualify for application to nanotechnology. Then considering that neurotechnology – a rapidly growing field in the intersection between nano and biotechnology – can affect moral capacities of the brain, she suggests that ethics itself may be affected by nanotechnology. And she leaves as open the question of a co-construction of ethics and bio nanotechnology. In short: new approaches must be considered, but which ones?

In Freitas (2005) it is affirmed that the society should be "able to muster the collective financial and moral courage to allow such extraordinarily powerful medicine to be deployed for human betterment, with due regard to essential ethical considerations".

At EU, there are some important concerns and discussion about the development of the integration of human beings and artificial (software/ hardware) entities. A funded project in this area (ETHICBOTS) was created to promote and to coordinate a multidisciplinary group of researchers into artificial intelligence, robotics, anthropology, moral philosophy, philosophy of science, psychology, and cognitive science, with the common purpose of identifying and analyzing techno-ethical issues concerned precisely with the integration of human beings and artificial (software/hardware) entities.

Next generations, in the perspective of nanomaterials utilization, will obtain many benefits in all these issues: health, economic, environmental, etc. By using this kind of advanced technology, with many more applications and much welcoming using, the conditions this technology offers will be very environmental friendly as well as already seen.

Actually, there are irreversible and long-term impact consequences for forthcoming generations and for the environment. However, there are risks, as well. But in truth there are incredible potential advantages. And the discussion is all over the place. That borders must be considered is, of course, evident and there has to be an achieved balance between the path found to the new dimension of knowledge and the contours specific for the problem. The rub of the question is exactly in what measure criticisms may have into account the technology potential.

Ethical Human Dignity

Despite the high feasibility for the economy and the environment, there are some considerations regarding the ethical, human dignity and moral borders on nanotechnology that should be taken into account.

It is interesting to have a look to the definition of human dignity and its connection to moral conveyed in the "opinion of the European Group on ethics in science and new technologies to the European Commission", 2005, when discussing the ethical aspects of information and communication technologies (ICT) implants in the human body. The group makes allusion to the EU's draft Treaty that Establishes a Constitution for Europe, stating that

> "Human dignity is inviolable. It must be respected and protected"

(Article II-61), and goes on to explain that

> "the dignity of the human person is not only a fundamental right in itself but constitutes the real basis of fundamental rights"

(Declaration concerning the explanations relating to the Charter of Fundamental Rights). This group declares yet that this explanation does not strictly define human dignity and that many writers have attempted to fill this gap. One such attempt suggests that human dignity is defined as follows:

> "the exalted moral status which every being of human origin uniquely possesses. Human dignity is a given reality, intrinsic to human substance, and not contingent upon any functional capacities which vary in degree. (…) The possession of human dignity carries certain immutable moral obligations. These include, concerning the treatment of all other human beings, the duty to preserve life, liberty, and the security of persons, and concerning animals and nature, responsibilities of stewardship."

The introduction of new devices to go further above the natural capabilities of the human being brings lots of alarms to scientific community. Desires, challenges but also concerns are always present… What are the consequences? To human beings, to life itself, to other living beings, to environment, to biodiversity, to progress, to the civilization as a whole...

Which are the consequences about the limit for the human being as human? Are humans creating a new being? Are humans prepared to get different, biologically and to become another thing? Are humans creating further "nano-digital devices" between humans that have access to these advanced technologies and humans that have not access? This discussion is recognized but probably the ethical discussion has not been enough promoted. And possibly no one is enough prepared to participate effectively and wisely on it. Much reflection, study and information is still needed.

Nanotechnology and Environment

In the long term, which are the consequences that nanotechnology has in the environment? Is it possible to predict that less damage in nature can be done if nanomaterials are considered instead of large equipment are used? Besides, once the need of inputs to produce these devices is lesser, probably there is an additional advantage on this perspective for nature preservation. But what is the reverse of the medal?

Let's have a look on the following situation:

- There is a growing consensus today that it is considered priority should be given to the development of alternative energies over the classical energies. It turns out that the classic energies price of production is cheaper than that of the alternative energies. But in the case of solar energy, produced with nano-photovoltaic panels, the price, though higher, is already comparable to, say, hydroelectric energy. In this case, this difference in costs can be compared with the environmental damage caused by the construction of dams. And even with the damage due to the pollution produced by the operation of the fuel oil plants. There is here an ethical problem in which one balances between paying a little more and preserving the environment or paying less and destroying it, in which the interests installed with the classical energies will hardly abdicate their positions without struggle, see (Ferreira et al., 2014).

Really, the implications and directions of nanotechnology need further discussion. The consequences of avoiding the discussion may be severe for society, considering rejection or fear of the effects. So it is pertinent to ask:

- Is there a need for a regulatory authority?
- Which kind of mechanisms are needed to regulate the nanotechnologies developments area?

ETHICS AND NANOTECHNOLOGY IN THE WORKPLACE

A theme urgently needing study and reflection concerns the potentially harmful effects on workers' health due to their exposure to nanoparticles in companies dealing with nanotechnology. Indeed, very little, to say nothing, is known about these harmful effects.

In similar situations the harmful effects on workers' health were only identified after long periods of activity, with the occurrence of real disasters in sanitary terms. One has to learn from history and try to avoid situations such as, for example, atomic energy in which, only relatively late, the harmful effects of radiation have been noticed. And only then were the necessary preventions taken, after the damage to health occurred in the workers of the plants and in the neighboring populations. And today humanity, in general, is not very confident on those preventions. However, atomic energy was initially used for war purposes which helps explain, but by no means excuse, this late reaction. The same is not true for nanotechnology, so it is natural that the attitude towards the problem is different.

Thus, in the absence of scientific knowledge on the potentially negative health effects of workplace exposure to nanoparticles, it is necessary to establish guidelines for decision making regarding risks and controls.

Identifying the ethical issues involved certainly helps with proper decision-making. It is therefore imperative for the decision-makers, particularly the employers, the workers, the investors, and the health authorities, to be aware of these issues. As the primary objective of occupational safety and health is the prevention of accidents and diseases in workers, the situations with the ethical implications that affect those most will be, see (Schulte & Salamanca-Buentello, 2007):

(a) Identification and communication of hazards and risks by scientists, authorities, and employers, both from the perspective of each of these groups and from a global perspective;
(b) Awareness and acceptance of risks by workers;
(c) Adequate choice and strict implementation of controls;
(d) Establishment of medical screening programs;
(e) Investment in toxicological research;
(f) Investment in control research.

Ethical issues involve the independent and impartial determination of risks, implacability, autonomy, justice, preservation of privacy, and promotion of respect for people. Their identification and implementation will make the options for decision-makers more evident. And company-wide discussions on the workplace hazards of nanotechnology can be reinforced by a special emphasis on small businesses and the adoption of a global perspective.

Ethical questions about nanotechnology in the workplace arise from the state of knowledge about the dangers of nanomaterials and the hazards and risks they may pose to workers' health. As this knowledge is still incipient, it is imperative to carry out a provisional assessment of the hazards and risks. Workers can exercise their autonomy only if the processes leading to hazard identification and risk assessment are transparent, comprehensible, and credible. Employers will be in accordance with the principles of autonomy, beneficence, non-maleficence, justice, privacy, and respect for persons, as far as, see (Schulte & Salamanca-Buentello, 2007):

(a) Portray hazards and risks without any omission;
(b) Address risks from a social security perspective;
(c) Communicate and dialogue with workers;
(d) Take the necessary measures to control risks so that they are understandable and accepted by workers to ensure their collaboration in their implementation.

Framework for Ethical Analysis in Nanotechnology

A methodology for conducting an organized and coherent discussion about ethics in this subject is to subdivide it by the following topics of analysis (see Mnyusiwalla et al., 2003): equity, privacy, security, environment, and metaphysical when discussing the relationship in the human-machine binary. About each one of them some considerations may be made. So, assuming these topics, and trying to go as far as possible in the analysis:

Equity

The classical dichotomy of developed and developing countries raises a new topic for discussion. Indeed, technology and development are closely correlated. Examples of great worries for developing countries are:

- People poverty reduction,
- Energy problem,
- Water problem,
- Health problem,
- Biodiversity maintenance.

Nanotechnology would help these countries to better conditions of population health and way of life. But it is costly to implement any project in this area and more often there are no financial resources available. Developing countries will be lesser benefited with these advances than developed countries. The main problem happens with the access to the benefits of technology considering the poorest and the richest people inside any country. Who may effectively reach the advances of science and take benefit from them? Anyway, it is true that poor people, whatever they are, may benefit, in some measure, from these advances. It may depend on the will of politicians and on the investments capacity (one way of overcoming countries' lack of investment capacity is their indebtedness which, as many current cases show, is nevertheless a generator of future and sometimes very serious financial problems)of some of these countries, for example. A result would be lower needs for energy and a cleaner energy production as well as other environmental benefits: a safer drug delivery, the improvement on health through for example better prevention, diagnosis and treatment.

Privacy

At this level, it is to consider the enormous innovations which can be reached, by improving surveillance devices or other devices that restrict human privacy. An individual may be facing situations of "invisible" microphones and cameras. This is a real problem if privacy is intended to be assured. Wireless monitoring provides healthcare providers and patients to be mobile and being able to exchange data when needed. But there is the reverse of the medal: privacy violation? It is true that wireless communication and new physiological sensors bring enormous implications in e-health. One point needed is about the necessary discussion on privacy and security.

Security

Surely new powerful weapons will be available very soon, as far as "invisible" microphones and cameras will be available. Will this ensure new ways of security or will be added to a potential "arsenal of bio-terrorism and techno-terrorism or even nano-terrorism?" (see Mnyusiwalla et al., 2003). Mnyusiwalla et al., 2003 also ask "who will regulate the direction of research in defensive and offensive military NT [nanotechnology]? How much transparency will be necessary in government and private NT initiatives to avoid misuses? There are also very interesting legal questions involving monitoring, ownership, and control of invisible objects".

Environment

This topic was already approached above in the section "nanotechnology and environment" due to extreme importance of the subject there considered: the production of electrical energy through nanophotovoltaic panels versus the production through classical means. Obviously this does not exhaust the subject. So, going on, new materials (fullerenes, carbon nanotubes) are now available. No one knows what will be the precise effects when this kind of nanomaterials goes into the environment (and also the effects on health). What are the dangers? Where are they put in the end of their mechanical lives? Will some medical devices be kept inside human body? What are the consequences? Or will be given another destiny to them? Is it possible to predict any kind of effects caused to the environment? Jacobstein (2006) refers that it is important to make an analysis on the risks associated to passive compounds in the less than 100 nanometer size range. There is the possibility, for example, of being introduced inadvertently in human bodies. There is in fact the concern with their ability to be inhaled, absorbed through the skin, or to pass through biological compartment barriers such as the blood brain barrier. This kind of dangers pose a real range of potential health and environmental risks that are associated to their potential toxicity or mutagenicity in their interactions with biological systems (note the analogy of this situation with the considered above in "ethics and nanotechnology in the workplace"). While the range of effects vary, most of the risks may be addressed by advanced industrial hygiene and environmental health practices and techniques that seek to characterize the specific risks, exposure patterns, and control methods and enforce them through a combination of practitioner education, industry self-regulation, monitoring and government regulation. This is an important emerging field in the environmental and health sciences, since most of the existing legislation on environmental, safety, and health risks may cover particulates, but do not take the change in physical and biological properties at the nanoscale into account. It is reasonable to assume that passive nanoscale particle risks, although potentially serious if not addressed, will be characterized and addressed systematically under new versions or extensions to existing occupational, industrial hygiene, environmental, and medical regulations.

Metaphysical

The incorporation of implants in human body, this is, the incorporation of artificial materials or machines into human systems may conduct to a concern around the human values and principles; will a human being be more than a human being? The borders are unknown. What kind of being and what kind of consequences. How this process may be controlled and accessed? One of the main issues is the "autonomy" of nano-bots, that is: robots whose components are at or near the scale of a nanometre

(10^{-9}meters), if they are able or not to move and to take decisions autonomously. Are they able to communicate each other and with the external environment? How can human beings ensure to "retire" them at the end of their "missions"? This is a very demanding question. The modification made on human bodies and living systems may create new realities. This is already a very real concern on the society, being in general the society often very skeptic about developments in this area. Catherine Larrère, see citation in Bensaude-Vincent (2010), discusses "the recent trends in nanoethics which anchor ethics in metaphysics or theology by emphasizing the emergence of new relations of men to nature and to God. It is defended that the moral issue raised by the project of enhancing human performances does not really lie in going beyond the boundary of human knowledge and condition. It is more a question of the moral choice underlying this new form of hubris".

Now in another although similar recording, nano-bots, or in general the "convergence" of bio-nano-info-neuro technologies, requires a new contract between science and society; society should not be seen as an afterthought only for making technology assessments, for social acceptability, etc. However, they should be really involved in the process; "converging technologies" - like syn-bio - should not be developed "inside the walls" of research labs only by highly specialized scientists, such challenges should be introduced requiring a cross-disciplinary approach, where project teams include anthropologists, sociologist, philosophers, and ethicists ... with a "precautionary principle" in mind...

Researchers in ETHICBOTS project in EU, which was finished in 2008, worked in order to identify techno-ethical case-studies on the basis of a state-of-the-art survey in emerging technologies for the integration of human and artificial entities; to identify and analyze techno-ethical issues concerned with the integration of human beings and artificial entities, by means of case-studies analysis; to establish a techno-ethically aware community of researchers, by promoting workshops, dissemination, training activities, as well as by the construction of an internet knowledge-base, on the subject of techno-ethical issues emerging from current investigations on the interaction between biological and artificial (software/hardware) entities; to generate inputs to EU for techno-ethical monitoring, warning, and opinion generation (see Capurro et al., 2005-2008).

Rodotà & Capurro (2005) show the limitations on ICT implants in the human body as deriving from an analysis of the principles contained in various legal instruments. They say that they should be assessed further by having regard to general principles and rules concerning the autonomy of individuals, which takes the shape of freedom to choose how to use one's body, "*I am the ruler of my own body*", freedom of choice as regards one's health, freedom from external controls and influence.

Conclusion

Ethical, legal and social implications in this context show the importance of nanotechnology to the society and the consequences that the development of nanotechnology possibly will bring to mankind in the future. The irreversibility and long-term impact of these new developments enforce to have in mind that it is necessary to take into account also the rights of future generations and the planet (like Hans Jonas' "Principle of responsibility").

Indeed, the future is there and came quick and loudly. Even before the right time. The innovations are going fast and sprout from everywhere. And although the endemic crisis, mainly economic but also social and politic, will bring some restrictions to the investments in health and research areas in many countries, new developments in this area will prevail and many discoveries will be made. The machine is very well assembled and perfectly lubricated: innovations are growing and development is there, as much as the will to go faster and further in many domains of scientific investigation. The research on this subject gets new results every day and new challenges have to be faced. Some work is going on this way and it is intended, in a forthcoming occasion, to analyze some data in order to detect and study the impact of the actual crisis on nanotechnology and mainly on applications to medicine and industry. The crisis may bring some restrictions, but there is a strong will to face the challenges and a new beginning is there.

Ethics concerning nanodevices and the borders of human beings considering nanotechnology in general and applied to medicine (where these concerns assume usually a more dramatic aspect) also need a new research to be developed next. This debate of the borders of the human being regarding the application of nanotechnology may be also enlarged in order to have a study of the impact into telemedicine and e-health.

The ethical issues related with the themes "nanotechnology and environment" and "nanotechnology in the workplace" are extremely critical, difficult to approach subjects whose development is hard to predict, due to the lack of knowledge of the situations. It can only be filled in the future after observing the practical situations. However, it all requires the try to create credible scenarios that can frame future decision-making processes. It is also urgent, in both these cases and the above referred, to think about what kind of authorities, possibly with a regulatory vocation, will need to be created to supervise the activities that will appear in the field of nanotechnology.

REFERENCES

Ackerman, M., Craft, R., Ferrante, F., Kratz, M., Mandil, S. & Sapci, H. (2002). Telemedicine technology. State-of-the-art telemedicine/ telehealth symposium: An international perspective. *Telemedicine Journal and e-health*, *8* (1), 71–78.

Allhoff, F. (2009). The coming era of nanomedicine. *The American Journal of Bioethics*, *9*(10), 3-11.

Allhoff, F. (2009). Response to open peer commentaries on "The coming era of nanomedicine". *The American Journal of Bioethics*, *9*(10), W1-W2.

Bensaude-Vincent, B. (2010). Opening the Field of Nanoethics: Editorial Introduction to the Special Issue on Bionanoethics, II. *HYLE – International Journal for Philosophy of Chemistry*, *16* (1), 1-2.

Capurro, R. (2006). *Reflections on ethical aspects of nanomedicine*. In http://www.capurro.de/nanoethics.html.

Capurro, Rafael., Tamburrini, Guglielmo., Weber, Jutta., (editors) Luca, Botturi., Edoardo, Datteri., Francesco, Donnarumma., Mark, Gasson., Satinder, Gill., Alessandro, Giordani., Cecilia, Laschi., Federica, Lucivero., Pericle, Salvini., Matteo, Santoro., Guglielmo, Tamburrini., Jutta, Weber. & Kevin, Warwick. (Authors) (2005-2008). Deliverable D5 – Ethical issues in Brain Computer Interface Technologies. Ethicbots Consortium, c/o University "Federico II" of Naples, *Ethicbots project - Emerging Technoethics of Human Interaction with Communication, Bionic, and Robotic systems Coordination Action - FP6 - Science and Society*, in http://ethicbots.na.infn.it/restricted/doc/D5.pdf and http://ethicbots.na.infn.it/goals.php.

Cattaneo, A. G., Gornati, R., Sabbioni, E., Chiriva-Internati, M., Cobos, E., Jenkins. M. R. & Bernardini, G. (2010). Nanotechnology and human health: Risks and benefits. *Journal of Applied Toxicology*, *30*(8), 730-744.

Cavalcanti, A. (2009). *Nanorobot Invention and Linux: The Open Technology Factor. An Open Letter to UNO General Secretary*, Melbourne, Australia, in http://www.cannxs.org/open.pdf.

Chavaglia, J. & Filipe, J. A. (unpublished draft). *Nanotechnology and Processes: Nano-Photovoltaic panels as an innovation in energy market.*

Chavaglia, J. & Filipe, J. A. (2011). Photovoltaic solar energy contribution for Amazon's companies. *International Journal of Academic Research*, *3* (4), Part I, 225-231.

Chavaglia, J., Filipe, J. A., Ferreira, M. A. M. & Coelho, M. (2012). Nanotechnology and Processes: Nano-Photovoltaic Panels as an innovation in Energy Market. *International Journal of Latest Trends in Finance and Economic Sciences*, *2*(3), 197-205.

Cruz-Cunha, M. M., Tavares, A. & Simoes, R. (2010). *Handbook of research on developments in e-health and telemedicine: Technological and social perspectives*. Hershey, PA: IGI Global.

Euroresidentes. (2011). *Introdução à Nanotecnologia: O que é Nanotecnologia*. Assessed in 26 of January 2011. Available in www.euroresidentes.com. [*Introduction to Nanotechnology: What is Nanotechnology*]

Ferreira, M. A. M., Filipe, J. A. & Chavaglia, J. (2014). Nanotechnology and Processes: The Nano-Photovoltaic Panels. *Advanced Materials Research*, *837*, 694-698. DOI: 10.4028/www.scientific.net/ AMR.837. 694.

Ferreira, M. A. M., Filipe, J. A., Coelho, M. & Chavaglia, J. (2017). Nanotechnology applications in industry and medicine. *Acta Scientiae et Intellectus*, *3*(2), 31-50.

Filipe, J. A., Chavaglia, J. & Ferreira, M. A. M. (2015). A Note on Nano-Photovoltaic Panels Emergence in Energy Market. *International Journal of Latest Trends in Finance and Economic Sciences*, *5*(1), 841-845.

Filipe, J. A. (2015). Nanotechnology and Medicine Improvement. *International Journal of Academic Research*, *7*(2), 32-37. DOI: 10.7813/2075-4124.2015/7-2/A.5.

Floridi, L. (1999). *Philosophy and Computing: An Introduction*. London - New York: Routledge, 1999.

Floridi, L. (2007). A look into the Future Impact of ICT on Our Lives, *The Information Society*, *23*(1), 59-64, 2007.

Floridi, L. (2008). Information Ethics, its Nature and Scope in *Moral Philosophy and Information Technology*, 40-65, edited by J. van den Hoven and J. Weckert. Cambridge: Cambridge University Press.

Floridi, L. (2010). *Information - A Very Short Introduction*. Oxford: Oxford University Press.

Floridi, L. (2011). *The Philosophy of Information*. Oxford: Oxford University Press.

Freitas, R. A. (2005). What is nanomedicine? *Nanomedicine: Nanotechnology, Biology and Medicine*, *1*(1), 2-9.

Freitas, R. A. (2007). Personal choice in the coming era of nanomedicine. In Patrick Lin et al. (Eds.), *Nanoethics: The Ethical and Social*

Implications of Nanotechnology, 161-172. New York, NY: John Wiley.

Hristozov, D. & Malsch, I. (2009). Hazards and risks of engineered nanoparticles for the environment and human health. *Sustainability*, *1*(4), 1161-1194.

Jacobstein, N. (2006). *Foresight Guidelines for Responsible Nanotechnology Development*. Draft version. Foresight Institute. In http://www.foresight.org/guidelines/current.html#Intro.

Lemos, R. T. S. & Kern, V. M. (2009). Technontologies, complexity, and hybrid interfaces. *Triple C*, *7*, 29-37.

Lindberg, S. (2005). *Hans Jonas's theory of Life in the face of Responsibility*. Available on http://www.estig.ipbeja.pt/~ac_direito/ Lindberg_HansJonas.pdf.

Kubik, T., Bogunia-Kubik, K. & Sugisaka, M. (2005). Nanotechnology on Duty in Medical Applications. *Current Pharmaceutical Biotechnology*, *6*, 17-33.

Malsch, I. (Ed.) (2011). *Nanotechnology and human health*. Utrecht: CRC Press.

Mcafee, A. & Brynjolfsson, E. (2008). Diferença competitiva. *Havard Business Review*.

McDonough III, E. F., Zach, M. H., Lin, H. E. & Berdrow, I. (2009). Integrando a inovação e o conhecimento na estratégia. *Revista Portuguesa e Brasileira de Gestão*. [Integrating innovation and knowledge into strategy. *Portuguese and Brazilian Journal of Management*].

Miyazaki, M. (2003). The Future of e-Health – Wired or not Wired. *Business Briefing: Hospital Eng. & Facilities Manag.* In http://www. touchbriefings.com/pdf/13/hosp031_r_miyazaki.PDF.

Mnyusiwalla, A., Daar, A. S. & Singer, P. A. (2003). 'Mind the gap': science and ethics in nanotechnology. *Nanotechnology*, *14*, R9–R13.

Moghimi, S. M., Hunter, A. C. & Murray, J. C. (2005). Nanomedicine: current status and future prospects. *The FASEB Journal.*, *19*, 311-330.

Moore, R. (2007). Medical nanotechnology: Evolving appropriate risk tools. *Medical Device Technology*, *18*(5), 36-37.

Navrátil, V. (2008). Future development of nanotechnology and human health. In *Contemporary Discourse on School and Health Investigation*. Brno: University of Brno.

Nogami, O. & Passos, C. R. (1999). *Princípios de economia.* São Paulo: Pioneira. [*Principles of economics*]

Roco, M. C. (2003). Nanotechnology: convergence with modern biology and medicine. *Current Opinion in Biotechnology*, *14*, 337–346.

Rodotà, S. & Capurro, R. (2005). *Ethical aspects of ICT implants in the human body*. Opinion of the European Group on Ethics in Science and New Technologies to the European Commission, nº 20, in http://ec.europa.eu/bepa/european-group-ethics/docs/avis20_en.pdf.

Schulte, P. A. & Salamanca-Buentello, F. (2007). Ethical and Scientific Issues of Nanotechnology in the Workplace. *Environ Health Perspect.*, *115*(1), 5-12.

Schumpeter, J. A. (1982). *Teoria do desenvolvimento economico.* São Paulo: Abril Cultural. [*Theory of economic development*]

Srivastava, D. & Atluri, S. A. (2002). Computational Nanotechnology: A Current Perspective. *Computer Modeling in Engineering & Sciences (CMES)*, *Tech Science Press*, vol. *3*, no. 5, 531-538.

Toothman, J. & Aldous, S. (2000). *How solar cells work.* Retrieved from http://science.howstuffworks.com/environmental/energy/solar-cell.htm.

Wiener, N. (1948). *Cybernetics - or Control and Communication in the Animal and the Machine.* Cambridge, Mass.: The MIT Press, 11-12.

Wiener, N. (1950). *The Human Uses of Human Beings: Cybernetics and Society.* Boston: Houghton Mifflin Co., 26-27.

Yan, H., Zhang, X., Shen, Z. & Seeman, N. C. (2002). A robust DNA mechanical device controlled by hybridization topology. *Nature*, *415*, 62-65.

http://www.understandingnano.com/nanotechnology-electronics.html. Accessed in 19.06.2017.

About the Editor

Dr. Paresh Chandra Ray, Professor of Chemistry at Jackson State University. His research group is heavily involved in nanotechnology research that include development of new class of nanomaterials, understating the fundamental physics, chemistry and material science behind the nanometer scale objects, exploring new chemical strategies for the imaging and multimodal therapy of cancer and drug resistance bacteria, designing multifunctional nanomaterial for developing nanomedicine, and enhancing our understanding of biomolecular interaction with nanosurface. He has been author/co-author for over 170 peer-reviewed publications in international and national renowned journals. His group's work has been widely cited with h-index 53. He holds several US pending patent. His research has been funded by National Science Foundation, National Institute of Health, Army Research Office, Army Research Lab, Universal Technology Corporation.

INDEX

A

B

C

D

E

I

L

M

N

O

P

Q

R

S